THE
HUMAN
COMPUTER

THE
HUMAN
COMPUTER

Mark Jeffery

LITTLE, BROWN AND COMPANY

A *Little, Brown* Book

First published in Great Britain in 1999
by Little, Brown and Company

Copyright © 1999 by Mark Jeffery

A CIP catalogue record for this book is
available from the British Library

ISBN 0 316 64843 4

Typeset in Baskerville by M Rules
Printed and bound in Great Britain by
Clays Ltd, St Ives plc

Little, Brown and Company (UK)
Brettenham House
Lancaster Place
London WC2E 7EN

CONTENTS

INTRODUCTION

Computers could be like humans in every respect. They could have the intelligence to understand Shakespeare's plays, Napoleon's military strategy, Einstein's theory of relativity. They could have the creativity to paint pictures and compose symphonies, design buildings and invent devices. They could be appreciative of humour and beauty, sensitive to criticism and compassion, motivated by curiosity and ego. And computers could be conscious in the same way as humans. Such computers do not currently exist. But they will.

To someone whose only experience of computers is of those found in offices and homes, these claims may seem far-fetched. Office computers used for writing documents, balancing accounts or processing data clearly bear no resemblance to humans; nor do home computers used for sending electronic mail, seeking information or playing games. These computers are no more than machines. They may sometimes behave in what seems to be an intelligent way, as when a computer running word-processing software corrects the spelling of a word, or when a computer running spreadsheet software balances this month's accounts. But such tasks, however impressive they may seem, can be performed according to mechanical procedures

that require no intelligence. Similarly, computers may sometimes behave in what seems to be an emotional way, appearing
to be joyful when they emit fanfares of sound when switched on,
or impatient as they beep repeatedly as users make mistakes. But
again, such sounds can be emitted according to mechanical
procedures. The computer's apparent joy and impatience represent no more than the projection of human emotions on to
the computer.

Even to someone with experience of computers vastly more
powerful than those found in offices and homes, claims that
computers might bear any resemblance to humans may seem
far-fetched. One example is the chess-playing computer Deep
Blue, which beat the world chess champion Garry Kasparov in
a widely publicised chess match in 1997. To many of those who
witnessed the victory, it seemed somehow hollow. Deep Blue
seemed detached from the victory, indeed, from the entire
match, an impression encouraged by the fact that one of the
computer's programmers, rather than the computer itself, actually moved the pieces on the chess board. Furthermore it was
the computer's programmers, rather than the computer itself,
who were seen to celebrate the victory. But there was a deeper
sense in which Deep Blue was detached from the match. The
computer was able to assess moves at an astonishing rate, but,
once again, it was merely performing mechanical procedures. It
could not appreciate the skill either of its own or of its opponent's moves, or feel the excitement of the occasion, or feel the
fear of defeat or the thrill of victory. The computer had no
more understanding of the chess match it was being used to win
than a sewing machine has of the dress it is being used to make.
If Deep Blue was intelligent, its intelligence was utterly unlike
that of humans.

Claims that computers might resemble humans may even
seem far-fetched to someone familiar with the computers found
in artificial intelligence research laboratories, where small robots
trundle around artificial environments, performing simple tasks
such as manipulating blocks. The behaviour of these robots is

interesting and certainly amusing, but the computers that control them can scarcely be credited with intelligence any more advanced than that of insects.

The uncertain progress towards the creation of computers as intelligent as humans, let alone creative, emotional and conscious in the same way as humans, has fuelled debate on what has become one of the most complex and controversial issues of our time. What makes it so complex is that it spans such disparate fields as mathematics, physics, neurology, psychology and philosophy. What makes it so controversial is that it encompasses such fundamental questions as whether humans have souls, whether mind and matter are distinct, whether consciousness can be explained, whether humans are unique in the universe.

On one side of the argument are those who hold that the uncertain progress of artificial intelligence research reflects only the difficulty of the research, and that, as technology advances, computers that are like humans will eventually be realised. Others argue that the uncertain progress reflects fundamental differences between humans and computers, and that, no matter how far technology advances, computers will never be like humans. These arguments are not based solely on the progress of artificial intelligence. Indeed, the arguments are as diverse in nature as the mathematicians, physicists, neurologists, psychologists and philosophers who propose them.

Some claim that computers are mere manipulators of information, unable to understand the information they manipulate. The philosopher John Searle devised the analogy of the Chinese room to illustrate such an argument. The Chinese room has no windows and no doors. Its only opening is a narrow slit through which questions written on a slip of paper in Chinese may be passed into the room. Shortly after a question is passed through the slit, an intelligent answer, again written on a slip of paper in Chinese, is passed back out of the room. The room is occupied by a human who does not understand the Chinese language. Whenever he receives a question, he analyses the Chinese characters according to a comprehensive set of instructions written in a language he does understand. For every possible question, the instructions indicate whereabouts on the countless shelves in the room a slip of paper giving an appropriate answer may be found. By following the instructions, the human appears to be intelligent enough to be able to provide an intelligent answer to any question. But because he does not understand Chinese, he has no understanding whatsoever either of the questions or the answers. Searle compares the Chinese room to a computer. He suggests that by following the instructions provided by its programmer, a computer may appear to be intelligent, but in fact it has no grasp at all of the information it processes.

Other arguments proposing a fundamental difference between humans and computers are based on a distinction between mind and matter. This distinction was expounded by the philosopher René Descartes in the seventeenth century, but since then support for it has slowly diminished. Scientists in later centuries were so successful in their attempts to describe the world in terms of matter that the concept of mind as a separate entity came to seem superfluous. However, interest in the distinction between mind and matter has recently been revived. The mathematician David Chalmers argues that the human mind cannot be explained by the laws of physics alone, and suggests that an entirely new set of laws, somewhere between the realms of physics and psychology, will be needed to explain the relationship between mind and matter. If Chalmers is right, there seems little prospect of creating computers that are like humans until this new set of laws is discovered. The philosopher Colin McGinn goes further. He argues that humans may be no more capable of understanding the relationship between mind and matter than an insect is of understanding Einstein's theory of relativity. If McGinn is right, the prospects for the creation of computers that are like humans seem bleak indeed.

Other arguments advancing a fundamental difference between humans and computers have a more scientific flavour.

The mathematician and physicist Roger Penrose argues that computers are unable to understand certain mathematical concepts. He expands on this claim to come to the same conclusion as Searle with his Chinese room, that computers are unable to understand the information they manipulate. Penrose provides a speculative explanation of how humans can understand the mathematical concepts that computers cannot understand, suggesting that the behaviour of microscopic tunnels of protein in the human brain may be governed by as yet undiscovered laws of physics fundamentally different from those governing the behaviour of computers.

I consider each of these arguments for a fundamental difference between humans and computers to be flawed. I reject Searle's argument that computers could not be made to understand the information they manipulate. The analogy of the Chinese room is confused. Searle views a computer as separate from the instructions provided by its programmer, in the same way as the human in the Chinese room is separate from the instructions with which he is provided. In fact, a computer is defined by its instructions; indeed, a computer is incapable of manipulating information, let alone understanding information, without instructions. The computer must be compared not to the human inside the Chinese room, but to the room as a whole, instructions included, and at this point Searle's analogy breaks down. A computer cannot be considered to 'contain' an intelligent being in the same way as the Chinese room contains the human, any more than the human brain can be considered in this way. As I will argue in the chapter on consciousness, any conception of the human brain that considers it to 'contain' an intelligent being is misconceived. Searle further seems to consider that the only way in which a computer can manipulate information is by following comprehensive instructions that specify an appropriate response to every possible stimulus (or, to persist with the analogy of the Chinese room, an appropriate answer to every possible question). As I will argue in the chapter on intelligence, a computer programmed in such a way would indeed be incapable of understanding. But computers could be made to manipulate information in far more flexible ways than this.

I reject any distinction between mind and matter as at best unnecessary, and at worst quite misleading. I do not consider humans to be separate from the physical world, but part of it, subject to the same physical laws as a mountain or a mite. These physical laws may seem too simple or too definite to give rise to the complexities and uncertainties of human intelligence and consciousness, but instances of simple physical laws giving rise to complex behaviour are common. Consider the physical laws that govern the motion of molecules of air and water in the earth's atmosphere. These laws, known as the laws of thermo-dynamics, are both simple and definite. Simple rules relating to pressure, volume and temperature explain how pockets of hot air rise through cold air, and how winds blow around regions of low and high pressure; simple rules governing the evaporation and condensation of water explain how clouds form and how rain falls. Yet the motion of air and water in the atmosphere is both complex, as shown by even the most cursory observation of the weather, and uncertain, as any attempt at weather forecasting will prove.

I reject Penrose's suggestion that new laws of physics are required to explain the behaviour of the human brain. In the chapter on consciousness, I will counter the arguments put forward by Penrose and others that consciousness derives from certain properties of the laws of physics that apply to humans but not to computers.

However, my assertion that computers could be like humans will not be negative, based on the rejection of arguments for a fundamental difference between humans and computers. Instead, my case will be positive, based on an account of how computers could be made to be like humans. As a necessary prelude to this account, I will consider what it means to be human, what is the nature of human intelligence, creativity, emotion and consciousness. I will describe how, in principle, computers that are intelligent, creative, emotional and conscious might be created. Finally, I will argue that such computers would be no mere lifeless imitations of humanity, but truly like humans in almost every respect.

I will not describe in every detail how computers could be made to be like humans in practice. Artificial intelligence researchers have worked for a long time, and will continue to do so, on the operational problems involved. Instead, I will present an overview of how computers might be made to be like humans in principle. I will suggest solutions to practical problems only to illustrate that these problems are not insurmountable. Further, I will not *prove* that computers could be like humans. Even if I were able to present the reader with a computer that appears to be intelligent, creative, emotional and conscious in the same way as humans, I could not *prove* that it is truly like a human. Instead, I will present a consistent and convincing way of looking at humans and computers to support the assertion that computers could be like humans in every respect.

This book is about humans and computers, about intelligence, creativity, emotion and consciousness. It is about the intriguing possibility that the computers of the future will be as worthy of the epithet 'human' as their human creators.

1
THE INTELLIGENT COMPUTER

Intelligence

Intelligence is commonly ascribed only to people with extra-ordinary talents, such as the ability to speak fluently in several languages, write a sophisticated critique of a work of art, or solve complex mathematical equations. For psychologists, intelligence has a wider meaning, encompassing such ordinary talents as the capacity to recognise a face, read a newspaper, conduct a conversation, play a sport, or even just walk down the street. Because we exercise these abilities effortlessly every day, we tend to take them for granted, and hesitate to associate them with intelligence. But the processes involved – both ordinary and extraordinary – turn out to be much the same. This chapter describes these abilities, how they are realised in the human brain, and how they could be reproduced on a computer.

Fundamental to human intelligence is our ability to learn information. Learning in this context is not confined to formal instruction in schools and colleges, nor is information in this context confined to that found in textbooks and encyclopaedias. Whether we are studying Einstein's theory of relativity, reading an article in a fashion magazine, glancing at a clock to find out the time, or simply noticing that the cat has walked into the room, we are constantly acquiring knowledge.

The information we learn is flexible. We are able to manipulate it, apply it, and use it as the basis for further learning. Whether we are sitting an examination, choosing clothes to wear to a party, deciding when to leave the house to catch a train, or simply stroking the cat, we are constantly manipulating and applying our knowledge. Whether it is a child using her knowledge of the numbers up to ten as the basis for learning how to add, or a physics student using her knowledge of matrices and calculus as the basis for learning Einstein's theory of relativity, we are constantly using our knowledge as the basis for further learning. Moreover, our manipulation and application of information, and our subsequent use of it as a basis for further learning, are flexible, so that we can find new ways to manipulate and apply that information, new ways to learn. Flexibility is fundamental to human intelligence.

Intelligence is not an all-or-nothing phenomenon. Most people would agree that dogs are intelligent to a degree, but few would claim that dogs are as intelligent as humans. It was once conventional wisdom that humans were fundamentally different from other animals, and in particular that human intelligence was fundamentally different from that of other animals. Since Darwin, Wallace and others developed theories of evolution, according to which humans and animals share the same origin, this position has become increasingly difficult to defend. Humans are better able to acquire, manipulate and apply knowledge than other animals, but an evolutionary perspective reveals these differences to be no more than differences of degree. If humans are more intelligent than chimpanzees, which are more intelligent than dogs, which are more intelligent than mice, and so on, then perhaps there is a place for computers somewhere on the scale. Even if the computers in the artificial intelligence research laboratories of today only reach the level of cockroaches on the scale of intelligence, perhaps the computers of tomorrow or the day after will reach the level of mice, or of dogs, or of chimpanzees – or even of humans.

The Unintelligent Computer

Every publisher of computer software seems anxious to advertise the intelligence of its programs. Word-processing software checks your documents for spelling and grammatical mistakes, even for clumsiness of style. Educational software analyses your use of it and presents you with tuition tailored to your requirements. Games software provides hosts of intelligent baddies against whom to pit your wits.

Computers running such software may have been programmed to act intelligently. But they are not intelligent.

The principle of programming computers is simple. The programmer provides the computer with a precise set of instructions that tells it what to do in all possible circumstances. The computer then follows these instructions, doing exactly what the programmer has told it to do.

The instructions are an encoding of the programmer's knowledge of what the user of the computer might want to do, and how to do it. For example, a programmer of word-processing software knows that the user might want to underline words in a document. He also knows that a user-friendly way to allow the user to do this is to provide an underline button on the screen, which, when pressed, underlines the words the user has selected (see Figure 1). So the programmer provides the computer with instructions to draw a button labelled U (for Underline) on the screen. He also provides the computer with instructions to check frequently whether the mouse button is pressed, and, if so, to check whether the mouse pointer is over the U button, and, if so, to underline the words the user has selected. The computer must be provided with hundreds of precise instructions for each of these tasks.

The computer follows the instructions provided by the programmer in much the same way as a mechanical digger on a building site follows the instructions provided by the driver manipulating its controls. If the driver, by pulling the appropriate levers, instructs the digger to move forwards, the digger does

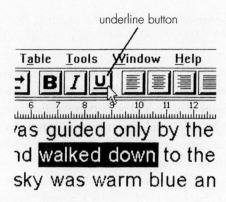

Figure 1. When the user presses the underline button, the selected words are underlined.

so without question. Similarly, if the programmer instructs the computer to draw a button labelled U̲ on the screen, the computer does so without question. Neither the digger nor the computer understands the instructions with which they are provided, they merely follow them. So if the driver instructs the digger to plough into a newly constructed brick wall, the digger will follow the instructions mindlessly, regardless of any damage to itself or to the wall. (When attempting to explain the accident to the foreman, the driver might claim that the digger has a mind of its own, but the foreman is unlikely to believe this.) Similarly, if the programmer instructs the computer to delete all the user's documents without the user's permission, the computer will follow the instructions mindlessly, without regard to the consequences. (Fortunately no programmer would intentionally provide a computer with such deleterious instructions, the notable exceptions being programmers of computer viruses.)

The only difference between the computer and the digger is one of time. The digger follows the instructions of the driver immediately, whereas the computer follows the instructions of the programmer some time after he has given them. The absence of the programmer when the computer is following his instructions can give rise to the eerie feeling that the computer is sentient. Anyone who has seen a pianola in operation, with its keys moving up and down to hammer out a tune without a pianist, will know this feeling. It is, of course, an illusion. The pianola is merely parroting the performance of a pianist whose notes have been recorded as perforations in the paper roll inside. If the pianola gives a virtuoso performance, the virtuosity is that of the pianist, not the pianola itself. Similarly, if a computer programmed as I have described acts intelligently, the intelligence is that of the programmer, not the computer. The programmer has decided, using his own intelligence, what would be the most intelligent way for the computer to act under all possible circumstances, and has provided the computer with precise instructions to act in this way. The computer has no intelligence of its own.

An illustration of the unintelligence of such a computer is provided by the computer's behaviour when the set of instructions provided by the programmer is incomplete or incorrect. Consider the software that controls the movement of the mouse pointer in a word-processor (see Figure 2). Suppose the programmer of this software fails to provide the computer with instructions for dealing with the user attempting to move the mouse pointer off the edge of the screen. This is an inadvisable thing for the user to do; once the mouse pointer has disappeared, she will not know where it is, and so will not know how to move it back on to the screen. She is liable to lose the mouse pointer completely. In the absence of specific instructions to the contrary, the computer will make no attempt to prevent this from happening, with frustrating consequences for the hapless user. Of course, a good programmer will foresee and forestall this eventuality by instructing the computer to check, whenever the user moves the mouse, whether moving the mouse pointer in response would cause it to disappear off the edge of the screen. He will instruct the computer to proceed with the moving of the mouse pointer only if this would not cause it to vanish, and otherwise to leave it visible at the edge of the screen. So it is the

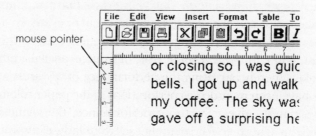

Figure 2. The user can move the mouse pointer anywhere on the screen, but cannot move it off the edge of the screen.

programmer who must employ his intelligence to foresee and forestall this problem, because the computer exercises no intelligence of its own.

If the programmer makes a mistake in the set of instructions with which he provides the computer, the consequences for the user may be even worse. Suppose a gardener were given the following instructions in a note left by a householder: 'Please mow the lawn and put the cut grass on the dog at the back of the garden.' The householder has made a mistake in the instructions. What she really wanted was for the gardener to put the cut grass on the compost heap at the back of the garden, but when she wrote the note she was distracted by her dog, and wrote 'dog' instead of 'compost heap'. The gardener, being an intelligent human, spots the mistake. Instead of subjecting the dog to a shower of grass, he guesses that the householder meant the compost heap, and puts the cut grass there.

Now, suppose a programmer of word-processing software makes a mistake in the instructions with which he provides the computer. The computer, being unintelligent, cannot notice the mistake, however obvious it may be to an intelligent human. It cannot guess what the programmer really intended. Instead, when it encounters the error, it will continue to follow nonsensical instructions mindlessly. The result is the word-processing equivalent of showering the dog with grass. The programmer's mistake may cause the user some confusion. For example, a mistake in the instructions followed by the computer when drawing the underline button on the screen may mean that the label U̲ fails to appear on the button, so that the user may not recognise it as the underline button. Worse, the programmer's mistake may prevent the user from using the software as intended. For example, a mistake in the instructions the computer follows when the underline button is pressed may actually prevent the user from underlining words. Worse still, the programmer's mistake may cause the user's documents to be lost. If the programmer fails to provide the computer with a correct and complete set of instructions, the computer fails to act intelligently.

If the programmer succeeds in providing the computer with a complete and correct set of instructions, the computer may act intelligently, but it is inflexible. When the programmer encodes his knowledge in a set of precise instructions, that knowledge becomes fixed. Suppose a programmer of word-processing software decides that most users would prefer the underline button to appear in the top-left corner of the screen. The computer cannot then reverse the programmer's decision and draw the button in the bottom-right corner instead. Of course, the programmer could provide the computer with instructions to allow the user herself to move the button to her preferred position on the screen, but it is the programmer's knowledge of what the user might want to do that prompts him to provide these instructions. The computer cannot change the instructions. It cannot manipulate the programmer's knowledge, cannot apply the programmer's knowledge in any way other than that encoded by the programmer.

Some software is promoted as able to 'learn'. Educational software analyses the user's knowledge and interests and presents different users with different information accordingly. Games software analyses the player's strategy, and optimises the computer's counter-strategy accordingly. Such software represents a step towards intelligence, because it allows the learning of information about the user, and because that information is flexible, in that it is manipulated and applied by the computer. But the ways in which the information is learnt, and the ways in which it is manipulated and applied, are not flexible, but fixed. The programmer has decided, using his own intelligence, what would be the most intelligent way for the computer to react to various classes of user. He has given the computer precise instructions as to what information to collect about users, how to analyse it to determine what class the user falls into, and how to react to users in each of the different classes. The computer cannot change the instructions. It cannot develop new ways of collecting information about users, or classifying or reacting to them.

Computers programmed as I have described may act intelligently, but they are not intelligent: the intelligence is that of the programmer, not the computer. For a programmer of word-processing software, this distinction is esoteric; it does not matter to him whether the intelligence is his own or that of the computer, as long as the computer reacts intelligently to the demands of the user. But the distinction is essential for the programmer of a computer that is to be intelligent in the same way as humans. There are real differences between the intelligence exhibited by a computer programmed as I have described and the intelligence exhibited by a human. The vicarious intelligence exhibited by the computer is fixed, inert, whereas the genuine intelligence exhibited by the human is flexible, active.

If a computer is to be intelligent in the same way as humans, it must be able to learn, manipulate and apply information in a flexible way. If the programmer uses his own intelligence to decide how the computer should act under all possible circumstances, such flexibility is precluded. Rather than providing the computer with instructions that encode his own knowledge, the programmer of the intelligent computer must give it instructions that allow it to acquire knowledge of its own. The newborn computer, like a newborn baby, will know next to nothing. But the intelligent computer, like the intelligent human, will acquire knowledge as it matures, knowledge that is truly its own.

It is tempting to assume that the human brain provides a perfect model of intelligence for the designer of an intelligent computer, but this assumption is as false as the one that the human body provides a perfect model for the designer of a robot. The structure of the human body, made of bones, muscles and tendons, is unlikely to be appropriate for a robot, made of metal, motors and wires. Considering the differences in the properties of metal and bone, and the differences in the operation of motors and muscles, it would be perversely anthropocentric to build a robot with metal struts moulded in exactly the same shape as human bones, and with electric motors mounted in exactly the same positions as human muscles. Similarly, the human brain is made of neurons and synapses, whereas a computer consists of capacitors and transistors. There is no reason to think that the structure of the human brain is appropriate for a computer, or that the human brain provides a perfect model of intelligence. In the same way some robots are able to lift greater weights than a human can, computers will eventually be designed whose intelligence surpasses that of humans.

Despite the differences between the neurons and synapses of the human brain and the capacitors and transistors of a computer, the human brain does provide invaluable hints for the designer of the intelligent computer. Indeed, computers could be made to acquire knowledge in much the same way as humans.

Sensation

The knowledge we acquire is derived from information about our environment. Philosophers have long argued over the existence of *a priori* knowledge, meaning knowledge that is independent of experience. For example, many philosophers have considered knowledge of logic and mathematics to be *a priori*, holding that a statement such as 'two plus two equals four' is necessarily true, so that we do not have to check it against information about our environment to ensure that all instances of putting two things together with another two things have, in our experience, resulted in four things. Whether the statement 'two plus two equals four' represents real knowledge or whether it is simply a consequence of the definitions of the words 'two', 'four', 'plus' and 'equals' is another question discussed by philosophers. Bertrand Russell has argued that, although *a priori* knowledge is independent of experience, it is elicited by experience. In other words, although the idea that two plus two equals four does not have to be continually verified against information about our environment, we learn it by observing that putting two things together with another two things always results in four things. So even our *a priori* knowledge is derived from information about our environment.

Information about our environment reaches our brains through our sense organs. Far too much information is available for us to be able to consider it all. Our eyes are capable of detecting the fall of individual photons on the retina, but our brains would have to be enormous if we were to remember each of these individual photon-falls. So, as we process the information from our sense organs, we distil the significant from the extraneous. As we resolve a scene from the plethora of photon-falls on to the retina, we extract the important aspects of the scene, such as the sabre-toothed tiger running at us with saliva dripping from its fangs, from the extraneous aspects, such as the colour of the leaves on the tree from which the sabre-toothed tiger has just emerged. This ability to distinguish between what is important and what is unimportant is fundamental to intelligence.

It would seem reasonable to suppose that the derivation of meaning from sensory information is a prerequisite for the distillation of the important information from the extraneous; in other words, that the processing of photon-falls on the retina into the idea of a sabre-toothed tiger precedes the selection of the sabre-toothed tiger as important. After all, it is difficult to imagine how we could select the sabre-toothed tiger as important *before* we have determined that what we are looking at is, indeed, a sabre-toothed tiger. However, this time-dependence is not as clear-cut as it first seems. The processing and selection of sensory information are concurrent and complementary.

We acquire knowledge, then, by collecting information about our environment through our sense organs, deriving meaning from the information by processing it, and reducing the information by selection on the basis of importance. Our sense organs, though, are not limited to the collection of information. By their very structure, they process information and select what reaches our brains. In other words, the first stage of our processing and selection of sensory information occurs not in our brains, but in our sense organs.

Our eyes detect what physicists call electromagnetic radiation. The electromagnetic spectrum includes radio waves, microwaves, infra-red radiation, ultra-violet radiation, X-rays and gamma rays, but the eye detects only the narrow band of visible light (see Figure 3). It is no accident that our eyes have evolved to detect this particular band. Sunlight is more intense in this band than in any other part of the electromagnetic spectrum – more than half of the energy of the sunlight reaching the surface of the earth falls within the band of visible light. Our eyes, then, have evolved to detect what electromagnetic radiation is available.

Nor is it an accident that human eyes have evolved to detect so narrow a band of the electromagnetic spectrum, rather than

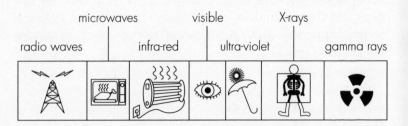

Figure 3. The electromagnetic spectrum.

a wider band including, say, infra-red and ultra-violet radiation. Some animals' eyes do detect a wider band: insects' eyes, for example, have evolved to detect near ultra-violet radiation as well as visible light; bees are thought to be able to recognise certain flowers more easily as a result. The primary reason for human eyes not having evolved in this way is evolutionary economy. The lenses in our eyes would have to be more complex in structure to focus infra-red and ultra-violet radiation as well as visible light. The evolutionary cost of developing such complex lenses outweighs the evolutionary benefit of the additional information in the infra-red and ultra-violet radiation. (For insects, the evolutionary cost-benefit analysis is different, because their compound eyes do not have focusing lenses.)

This economy has a secondary benefit. If our eyes had developed to detect infra-red and ultra-violet radiation as well as visible light, our brains would have had to have developed simultaneously to process the additional information in this radiation. Again, the evolutionary cost of developing the brain in this way outweighs the evolutionary benefit of the additional information in the infra-red and ultra-violet radiation. By detecting only visible light, our eyes reduce the amount of visual information reaching our brains by selection on the basis of importance.

Similarly, our ears reduce the amount of auditory information reaching our brains by failing to detect sounds that are particularly low-pitched or particularly high-pitched. The information in such low- and high-pitched sounds confers no particular evolutionary benefit, because these sounds rarely occur in our environment. Again, the primary reason for not detecting these sounds is evolutionary economy, but again, a secondary benefit is the reduction of the amount of auditory information reaching our brains by selection on the basis of importance.

As well as selecting what auditory information reaches our brains, our ears process that information by performing what electronic engineers would recognise as Fourier analysis. The output of a standard microphone is an electrical signal representing the intensity of the sound reaching the microphone at any particular instant. This intensity varies rapidly with time, too rapidly for our slow brains to cope with. Rather than convert the rapidly varying intensity directly into an electrical signal as a microphone does, the human ear transmits sounds into a spiral tube called the cochlea (see Figure 4). This tube is narrow at its base and wide at its apex, tapering gradually along its length. Sounds of low pitch cause sensors towards the wide end to vibrate, and sounds of high pitch cause sensors towards the narrow end to vibrate. These vibrations are transmitted along nerves to the brain. So the brain receives auditory information in the form of a complete analysis of the constituent pitches of sound reaching the ear. If an electronic engineer wanted to perform the same Fourier analysis on the output of a microphone, he would have to build an electronic circuit, or feed the output into a computer to perform the analysis mathematically. By their very structure, then, our ears perform preliminary processing of auditory information before it reaches our brains.

Just as the knowledge we acquire is derived from information about our environment, the knowledge acquired by the intelligent computer would be derived from information about its environment. The designer of the intelligent computer must

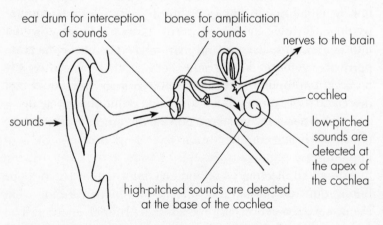

Figure 4. The human ear transmits sounds into the cochlea.

provide it with sensors for the collection of this information. Scientists and engineers have invented devices for the detection of an enormous range of information, not only of the visible light, audible sound, movement, contact, pressure, temperature, taste and smell sensed by humans, but also of radio waves, microwaves, infra-red radiation, ultra-violet radiation, X-rays, gamma rays, vibrations, ultrasonic sound, air pressure, air humidity, and so on. Any of these devices could be connected to the intelligent computer as sense organs, regardless of whether they correspond to human sense organs. For example, the intelligent computer could be provided with infra-red sensors instead of, or as well as, visible light sensors. This would be particularly useful for seeing people in the dark, because warm-blooded animals emit infra-red radiation continuously. (Indeed, rattlesnakes have evolved simple infra-red sensors that help them to locate warm-blooded prey.) The designer of the intelligent computer might be tempted to provide it with all the sensors listed above, and a few more besides.

But there are good reasons for avoiding this temptation. As discussed, the primary reason why human sense organs provide the brain with selected information about the environment, rather than all the information available, is evolutionary economy. The same demands of economy would apply to the intelligent computer. The designer of the intelligent computer would want value for money from his creation, and so would impose the same demands of economy on the intelligent computer as evolution has imposed on humans. Providing the intelligent computer with more and more sensors would cost more and more money, but would yield diminishing benefits.

There is also a technical reason for avoiding the temptation to ply the intelligent computer with sensors. A secondary benefit of human sense organs providing the brain with selected information about the environment is the discriminate reduction of the amount of sensory information reaching our brains. If humans were to evolve eyes able to see infra-red and ultra-violet radiation as well as visible light, we would also have to evolve brains capable of processing this extra information about our environment. Similarly, if the intelligent computer were plied with sensors, it might not be capable of processing all the information it would gather. Such information overload would be particularly problematic in earlier intelligent computers, which would inevitably be less powerful than later models.

Given that it must not be provided with too many sensors, the designer of the intelligent computer must decide what sensors to provide it with. The simple solution would be to provide it with sensors that detect the same information as human sense organs. It is not important whether the intelligent computer's sensors operate in the same way as human sense organs, or have the same structure as human sense organs. For example, the video camera and the human eye operate in different ways, and their structures are quite different, but they detect the same band of visible light, are sensitive to light of about the same intensity (though the human eye is superior in detecting low-intensity light), and have roughly the same resolution (though the human eye's is superior).

Providing the intelligent computer with sensors equivalent to human sense organs would certainly facilitate communication between the intelligent computer and humans. Human speech is well matched to human hearing. The range of pitches produced by our vocal cords is well within that detected by our hearing, and the volume at which we speak most comfortably is just right for the ideal volume at which we listen when speaker and listener are a short distance apart. Similarly, human writing is well matched to human vision. The text printed on this page is about the right size for most humans to read it comfortably when holding it a short distance from their eyes. If communication between humans and intelligent computers is to adopt the traditional forms of speech and writing, intelligent computers will require sensors similar to human sense organs. Existing forms of communication between humans and unintelligent computers, such as humans typing text using a keyboard and the computer displaying this on a screen, would be highly unsatisfactory for successful communication between humans and intelligent computers. Anyone who has used a text telephone, where communication between humans sitting at different computers proceeds in just such a manner, will understand how inadequate this is compared to using a voice telephone. Not only is typing slow and tedious, but the considerable amount of information communicated in tone of voice and manner of speech is lost when words are typed rather than spoken.

Providing the intelligent computer with sensors equivalent to human sense organs would facilitate communication in another, more subtle way. If the intelligent computer had sensors that differed significantly from human sense organs, its view of the world would differ from ours. Inevitably, communication between beings with similar perspectives on the world would be easier, though perhaps less instructive, than communication between beings with entirely different perspectives.

So the intelligent computer's sensors would perform preliminary processing and selection of sensory information, just as human sense organs do. For example, the intelligent computer's auditory sense organs would include an electronic circuit to perform Fourier analysis of sound, just like the cochlea in the human ear. It might be argued that by including the Fourier analysis circuit, the designer of the intelligent computer is using his own intelligence to determine how the auditory information reaching the computer should be processed, so that at least part of the intelligence exhibited by the computer is that of the designer, not the computer.

However, it might equally be argued that by including the cochlea in the human ear, evolution has used *its* intelligence to determine how the auditory information reaching humans should be processed, so that part of the intelligence exhibited by humans is in fact that of evolution. Both arguments are false. There is no intelligence in a Fourier analysis circuit, just as there is no intelligence in the cochlea. Evolution has provided humans with cochleas not through intelligence (evolution has no intelligence), but through billions of years' worth of selected accidents. So the designer of the intelligent computer may provide it with the Fourier analysis circuit without compromising its intelligence, just as evolution has provided humans with cochleas without compromising our intelligence.

Perception I

The information about our environment collected through our sense organs is further processed in our brains. Our eyes, for example, provide us with a huge amount of information about our environment, and our brains process this information at an impressive rate. We need only glance out of a train window to resolve the scene into telegraph poles, trees, fences, fields, hills and clouds. Such feats of processing are achieved through the interconnections of neurons in the brain.

The human body consists of billions of building blocks called cells. Some of these cells, called neurons, are specially adapted for the transmission of signals. Sensory neurons transmit signals from the sense organs to the spinal cord and the brain. Motor neurons transmit signals from the spinal cord and the brain to the muscles. Consider a reflex reaction such as the involuntary withdrawal of your hand when you accidentally hit your thumb with a hammer (see Figure 5). Sensory neurons extending from your thumb to your spinal cord each collect signals from a number of pain receptors. On the impact of the hammer, these signals increase the activation of the sensory neurons above a certain threshold, causing them to transmit a signal to your spinal cord, and in turn motor neurons extending from your spinal cord to your arm each collect signals from a number of these sensory neurons. On the impact of the hammer, these sensory neuron signals increase the activation of the motor neurons above a certain threshold, causing them to transmit a further signal to your arm, where the motor neurons are connected to muscles. On the impact of the hammer, the signals in the motor neurons cause these muscles to contract, achieving the withdrawal of your hand.

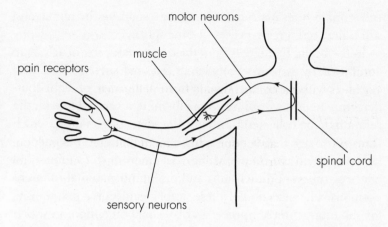

Figure 5. When you accidentally hit your thumb with a hammer, signals are transmitted from your thumb to your spinal cord along sensory neurons, and from your spinal cord to your arm along motor neurons, achieving the withdrawal of your hand.

In addition to sensory and motor neurons, there are some hundred billion neurons in the brain connected only to each other. These neurons transmit signals in the same way as sensory and motor neurons, but their purpose is not the transmission of information, but the processing of it. It is through the interconnections of these neurons that the brain processes the information collected through our eyes to achieve visual perception.

Light enters the eye through the pupil and is focused by a lens on to the retina, where it is detected by receptors. Some of these receptors detect red light, some green light, and some blue light, while others are not sensitive to colour. The information collected by the receptors is transmitted to the brain along sensory neurons in the optic nerve. Each of these neurons collects information from a number of receptors in the same small region of the retina, encoding differences between the light falling on the centre of the region and the light falling on the edge of the region. For example, some neurons transmit signals if the centre of the region is light and edge is dark, indicating a light spot on the retina. Other neurons indicate dark spots on the retina, and still others red, green, yellow or blue spots.

In the brain, the signals from these spot-detecting neurons are combined by neurons that detect features such as lines and edges. For example, the signals from a number of light-spot-detecting neurons from receptors forming a vertical line on the retina may be collected by a light-line-detecting neuron, which transmits a signal indicating a light vertical line at that particular position on the retina. Millions of neurons detect lines and edges of different dimensions and orientations and at different positions. The detection of these simple features is a prerequisite for the more complex processes of visual perception, in which the spots, lines and edges are resolved into objects.

Simple feature detection can easily be reproduced on a computer. Computers can be programmed to perform any process that can be defined by a precise set of instructions. The detection of spots, lines and edges by neurons in the human brain are just such processes. The behaviour of a light-line-detecting neuron, for example, could be modelled using the following set of instructions:

> Consider in turn each of the light-spot-detecting neurons in a vertical line on the retina. If the light-spot-detecting neuron is transmitting a signal indicating a light spot, then increase the activation of the light-line-detecting neuron. Otherwise decrease its activation. If, once each of the light-spot-detecting neurons in the vertical line has been considered, the activation of the light-line-detecting neuron exceeds a certain threshold, then transmit a signal along the light-line-detecting neuron to indicate a light vertical line.

For a computer to understand these instructions, they would have to be written in a computer language rather than in English, but the translation would be a simple task for a computer programmer.

As suggested earlier, it is important not to assume that the human brain provides a perfect model of intelligence. Simply duplicating the behaviour of the neurons in the human brain may not be the best way to reproduce visual perception on a computer. Indeed, most computer systems for processing images currently employ mathematical techniques such as differentiation and Fourier analysis to achieve simple feature detection, particularly edge detection. These mathematical techniques may, in time, be abandoned for an approach based on the neural model. The essential point for the discussion of the intelligence of computers, though, is that simple feature detection can be reproduced on a computer in principle, independent of how it is achieved in practice.

Beyond simple feature detection, the processing of visual information in the human brain becomes more complex. Neurologists do not yet understand how the higher processes of visual perception are achieved through the interconnections of neurons in the brain. Despite this lack of understanding of *how* these processes are achieved, psychologists have gained a good understanding of *what* the processes achieve.

Three processes are central to visual perception: distance determination is the process whereby two-dimensional features are interpreted as representing objects at particular positions in three-dimensional space; scene segmentation is the process whereby features are interpreted as representing discrete objects; and object recognition is the process whereby these objects are identified as telegraph poles, trees, fences, fields, hills or clouds.

We use a number of cues to determine distances and so reconstruct the three-dimensional world from the two-dimensional images on our retinas. If a near object and a far object overlap in our field of vision, the near object always obscures the far object (see Figure 6), so the interposition of objects provides an indication of how far away the objects are relative to each other. Further, near objects appear at very different positions in the fields of vision of the left and right eyes, whereas far objects appear at only slightly different positions (see Figure 7), so the

Figure 6. The tree on the right is nearer to the observer than the house, and so obscures it. The tree on the left is farther from the observer than the house, and so is obscured by it.

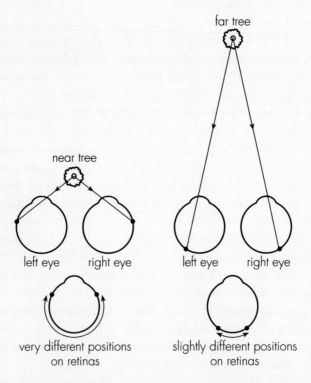

Figure 7. The image of the near tree forms at very different positions on the retinas of the left and right eyes, whereas the image of the far tree forms at only slightly different positions.

disparity in the position of an object in the fields of vision of the left and right eyes provides another indication of how far away it is. (Pictures and films that appear to be three-dimensional when seen through red and green glasses achieve the effect by presenting different images to the left and the right eyes in such a way that the viewer interprets the disparities in the images as representing depth.) Finally, near objects move across our field of vision more quickly than far objects. As seen from the window of a train, telegraph poles along the railway are a scarcely discernible blur, the trees beyond the telegraph poles are seen to move quickly, the fields beyond the trees move more slowly, and distant hills hardly seem to move at all. So the speed at which an object moves across our field of vision provides yet another indication of its distance from us. (Psychologists have found that we need not be observing from a rapidly moving position – we need only shift our heads very slightly – for this cue to contribute to our determination of the distances of objects.)

In addition to these simple cues for distance determination, we use a number of cues that require prior knowledge of the objects we see. Near objects appear to be large, whereas far objects appear to be small, so a comparison of the expected and apparent size of an object provides an indication of how far away it is. Near objects appear to be coarsely textured, whereas far objects appear to be finely textured, so a comparison of the expected and apparent texture of an object provides another indication of its distance. Even such complex prior knowledge as where we expect to find objects can contribute to distance determination. We expect trees to be rooted in the ground, distant hills to be stretched along the horizon, and clouds to be suspended in the sky (sometimes, if we see a bank of clouds stretched along the horizon, we can mistake it for a line of hills). All these cues, and others besides, are combined in a flexible way to determine the distances of objects.

Further cues determine which features of the images on our retinas are connected, and so segment a scene into discrete objects. Features that are close together, or are similar in shape, texture or colour, or are moving at the same speed in the same direction, are likely to belong to the same object. A sharp turn in a line or an edge is likely to indicate a boundary between different objects. A line or edge that forms a closed loop is likely to form the boundary of a single object. Again, in addition to these simple cues, our prior knowledge of the usual size, shape, texture, colour and location of objects provides further indications of which features belong to which objects. All these cues, and others, are combined in a flexible way to segment a scene into discrete objects.

Unlike the processes of distance determination and scene segmentation, the process of object recognition is necessarily based on prior knowledge. Object recognition is the process by which we interpret a collection of features as representing an object, or one of a class of objects, that we have seen before. An object such as our next-door neighbour's dog Brutus looks different every time we see him. Sometimes we see Brutus from the front, sometimes from the rear, sometimes from the side, sometimes (if we are looking down on the neighbour's garden from our bedroom window) from above. Sometimes we see Brutus close up, so that he appears large in our field of vision, and sometimes we see him from a distance, so that he appears small. Brutus is sometimes walking, sometimes running, sometimes sitting, sometimes lying down. Despite the infinite variety of images of Brutus that form on our retinas, we are able to recognise him from different angles and distances, and in different poses. Similarly, objects in a class, such as the class of dogs, also vary infinitely. Some dogs are small, some large, some black, some brown, some fat, some thin. Despite the infinite variety of dogs, we are able to classify an animal as a dog even if we have never seen that particular dog before, or even if we have never seen a dog of that particular breed before.

Object recognition, then, is not simply a matter of remembering detailed images of objects, and matching the images on our retinas with the images in our memories, feature for feature.

We could not possibly remember a different Brutus image for every possible angle, distance and pose, nor a different dog image for every possible dog, including breeds we have never encountered before.

Instead, object recognition involves segmenting the object into subobjects using the segmentation process described above, then interpreting each subobject as a simple geometrical shape such as a cylinder. Seen from a distance, Brutus might be represented as having a horizontal cylinder for a body, four long thin vertical cylinders for legs, and a horizontal cylinder for a head (see Figure 8). The shapes of the subobjects, combined with the pattern of their relative positions and proportions, represent Brutus's geometry. This representation allows Brutus to be recognised from different angles and distances and in different poses: the features (the lines and the edges) that make up the image of Brutus vary with angle, distance and pose, but the shapes of the subobjects and the pattern of their relative positions and proportions remain the same. Geometry does not make the only contribution to object recognition. Simple cues such as texture and colour are essential for the recognition of Brutus as an individual dog, rather than just his membership of the class of dogs. Our prior knowledge of context also contributes to object recognition. We are likely to recognise the dog playing in our neighbour's garden as Brutus rather than

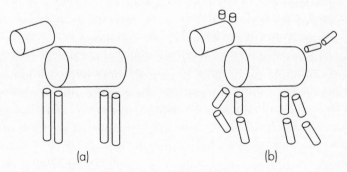

(a)   (b)

Figure 8. Cylinder representation of Brutus (a) from a distance and (b) from less of a distance.

any other dog, regardless of his actual features, because we are used to seeing Brutus in this context.

Descriptions of the processes of distance determination, scene segmentation and object recognition seem complete when these processes are considered in isolation. However, when these processes are considered as a whole, circular dependencies are revealed. It would seem reasonable to suppose that distance determination must precede object recognition, because the information provided by distance determination makes an important contribution to object recognition. But many of the cues for distance determination, such as our prior knowledge of the usual size, texture and location of objects, cannot operate before the object has been recognised. Further, it would seem reasonable to suppose that scene segmentation must precede object recognition, because the segmentation of a scene into objects seems a prerequisite for the recognition of those objects. But again, many of the cues for scene segmentation, such as prior knowledge of the usual size, shape, texture, colour and location of objects, cannot come into operation until the object has been recognised. It seems contradictory to say that object recognition cannot proceed before distance determination and scene segmentation are complete, yet distance determination and scene segmentation cannot proceed before object recognition is complete.

The resolution of this apparent contradiction lies in the realisation that we perform all three processes concurrently. Initially we use such cues as are available (for example, cues for distance determination and scene segmentation that are independent of object recognition) to arrive at a preliminary assessment of the objects before our eyes, which in turn provides further cues for distance determination, scene segmentation and object recognition. If the initial survey is accurate, the further cues reinforce the initial cues, allowing us to refine the assessment. If it is inaccurate, the further cues counter the initial cues, allowing us to revise the assessment. Eventually we arrive at progressively more accurate assessments of the scene through successive refinements and revisions. This view of visual perception is borne out by our experience of pictures of impossible objects (see Figure 9). We make an assessment of such a picture based on the part of the picture we see first. We are confirmed in this assessment until we allow our eyes to wander to a part of the picture incompatible with the initial survey. Our perception of the picture then suddenly changes, as we make an alternative assessment compatible with the new part of the picture. By passing our eyes back and forth across the picture we can cause our perception of it to flip again and again.

Our expectations make a significant contribution to our preliminary assessment of a scene. If we expect to see Brutus playing in our neighbour's garden, our preliminary assessment of the view from our bedroom window will be based on this expectation, in addition to such visual cues as are available. So if these visual cues indicate a small, dark object moving across the lawn, our preliminary assessment of the scene will be of Brutus playing on the lawn. As we arrive at progressively more accurate assessments of the scene, our expectations will be confirmed or contradicted. However, if we do no more than glance out of the window, or if it is night and our eyes are not adjusted to the darkness outside, there may be insufficient visual cues to arrive at these more accurate assessments. In such a case, our initial survey will constitute our entire perception of the scene.

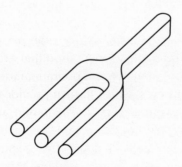

Figure 9. An impossible object. As our eyes wander between the bottom-left and the top-right parts of the picture, our perception flips between that of three cylindrical stumps and that of a two-pronged tuning fork.

Considering the extent to which our expectations, which are the products of a complex world-view, can influence our perception, we cannot persist in the misconception that our perception is objective. To an extent, we see what we expect to see, particularly when a paucity of other visual cues compels us to rely unduly on our expectations. Our experience of dreams provides a convincing demonstration of how compelling our non-sensory perception can be. Our dreams are generated in the virtual absence of real sensory information, so our perception in dreams consists almost entirely of constructions similar to those prompted by our expectations during our waking perception.

Dreams can nonetheless be vivid, to the extent that our experience of nightmares can be extremely unpleasant. However, when we wake up after a dream, we know that our dream experience was not real, and that it was not based on information from our sense organs (we may even be aware of this while still dreaming). What is disquieting about our waking perception is that we subconsciously combine cues based on the information from our sense organs with other cues based on prior knowledge, so we are generally unaware of the extent to which our perception is based on real sensory information, and the extent to which our expectations have filled the gaps. This influence of our expectations on our perception means that our world-views tend to be self-reinforcing, and our prejudices self-confirmatory.

Distance determination, scene segmentation and object recognition are more difficult to reproduce on a computer than simple feature detection. These more complex processes cannot be reproduced simply by modelling the behaviour of the human brain, because neurologists do not yet understand how they are achieved through the interconnections of neurons. This forces the designer of the intelligent computer to consider how best to achieve visual perception on a computer. The way in which the designer chooses to do this need not correspond to the way visual perception is achieved in the human brain. The brain is not a perfect model of intelligence, and what works well in the brain may not work well on a computer. What matters is whether or not the designer's chosen model for visual perception allows the computer to perceive its environment in the same way as humans.

Consider a model for the reproduction of distance determination, scene segmentation and object recognition on a computer. A computer with visual perception based on this model would, indeed, see things the same way as we do.

One of the cues for distance determination mentioned above is the disparity in the positions of features in the fields of vision of our left and right eyes. It would be relatively simple to reproduce the detection of this cue on a computer that has two video cameras for eyes. The computer could be programmed to compare the images from the left and right video cameras, matching features in one image with features in the other. It could then determine the disparity in the positions of each feature in the left and right fields of vision from the difference between its positions in the two images.

Another cue for distance determination already mentioned is the speed at which features move across our field of vision. Again, it would be relatively simple to reproduce the detection of this cue on a computer. Video cameras do not provide a continuously updated moving image, as our eyes do. Instead, they provide a succession of still images, typically fifty or sixty every second. The computer could be programmed to compare subsequent images to match features in one image with features in the previous image, determining the speed of each feature from the difference in its positions in the two images.

It would be relatively simple, then, to reproduce the detection of the cues for distance determination on a computer. However, the crux of distance determination is not the detection of the various cues, but their correlation. Whereas the detection of the cues would be innate, in that the programmer provides the computer with fixed instructions for the detection of the cues, the correlation of the cues would be learnt. The computer would note that features that appear at very different positions in the fields of vision of the left and right video cameras tend to move quickly across its field of vision, whereas features that appear at only slightly different positions tend to move slowly across its field of vision. It would learn further correlations between objects formed of the former features and objects formed of the latter features. It would discover that the former objects tend to obscure the latter, that the former objects tend to be large whereas the latter objects tend to be small, that the former objects tend to be coarsely textured whereas the latter objects tend to be finely textured, and so on. These correlations would provide the foundation for the computer's concept of distance. If the computer were provided with robotic arms, it would learn to recognise this concept of distance as determining whether it is able to reach out and touch an object, and how much it must extend its arms to do so. If the computer were provided with robotic legs, it would learn the further correlation between its concept of the distance of an object and how far it must walk to reach it.

Scene segmentation is similar to distance determination. Again, it would be relatively simple to reproduce the detection of the cues for scene segmentation on a computer. Again, the crux of scene segmentation is not the detection of the various cues, but their correlation, and though the detection of the cues would be innate, their correlation would be learnt. What we see as a toy train, the immature computer would see as a collection of features that are close together, similar in shape, texture and colour, bounded by an edge that forms a closed loop, and so on. Through its experience of this and other collections of features, the computer would learn that such collections of close, similar and bounded features tend to remain close, similar and bounded, always moving at the same speed in the same direction as each other. These correlations would provide the foundation for its concept of objects. Again, if the computer were provided with robotic arms, this concept would be reinforced by its reaching out and manipulating objects.

We are so used to thinking of the world as composed of discrete objects that we forget that this is only one way of seeing, and that the immature computer, like the immature human, might have to learn this way of seeing. We forget that our original concept of objects was of a set of visual features close together, similar in shape, texture and colour, moving at the same speed, in the same direction, and so on. In some ways, the infant's way of seeing is the more enlightened. Scientists in some disciplines no longer see the world in terms of discrete objects. Physicists think of subatomic particles not as discrete objects but in terms of continuous wavefunctions, and ecologists think less of discrete animals and plants than of whole ecosystems.

Object recognition is different from distance determination and scene segmentation. It involves not the correlation of different cues, but the development of a repertoire of ways of representing objects to allow their recognition. As discussed, one of the most important ways in which humans represent objects is as patterns of simple subobjects. If a computer is to represent objects in the same way, it must be able to develop a repertoire of simple geometrical shapes such as cylinders in order to represent the shapes of the subobjects, as well as a repertoire of common spatial patterns to represent their relative positions and proportions. In particular, if the computer is to recognise objects from any angle, it must be able to learn how the shapes and patterns of its repertoires appear from every possible angle. For example, it must be able to learn that a cylinder appears as a circle when seen end-on, and as a rectangle when seen side-on.

It seems contradictory that some of the processes of perception are learnt, but that we can learn only by deriving knowledge from information about our environment through perception; in other words, that learning is prerequisite to perception, yet perception is a prerequisite for learning. But this is not the contradiction it first seems, because perception and learning are not single, indivisible processes. The innate processes of perception provide the basis for initial learning, including the development of an understanding of simple processes of perception, which in turn provide the basis for further learning, including the learning of more complex processes of perception, and so on, incrementally.

To summarise, it would be relatively simple to reproduce on a computer the innate processes of visual perception, such as the detection of the cues for distance determination and scene segmentation. The real challenge is to reproduce the learnt processes of visual perception. The intelligent computer must be capable of learning the correlations between the various cues of distance determination and scene segmentation, understanding repertoires of simple geometrical shapes and common spatial patterns, and grasping how these shapes and patterns appear in different orientations. Whether these abilities could be reproduced on a computer is part of a larger discussion of whether computers could be made to learn at all.

Attention

Our ability to focus our attention on selected parts of the sensory information available to us makes a vital contribution to our distillation of the important from the extraneous. A fuller account of attention dealing with the issues of how we decide what to focus our attention on appears in the later chapters on emotions and consciousness, but a brief discussion is included here for the insights it affords into our selection of sensory information.

Our physical control of our sense organs makes an important contribution to our ability to focus our attention. This control is most evident in the case of our eyes. The small region at the centre of the retina is densely packed with receptors that are acute and sensitive to colour, so our brains receive considerably more visual information about what we are looking at directly than about the rest of our environment. By turning our heads and swivelling our eyes, we can focus images of different elements of the scene before us on to the centre of the retina, and so select which elements we see most accurately. This selection is so automatic that we are often unaware of it. When we read a book, we automatically scan the words on the pages. When we play a sport, we automatically scan the position of the ball and of the other players. When we face an unfamiliar scene, we automatically scan those elements of the scene that we consider to be the most important. Indeed, if our eyes remain still, the image detected by the receptors in our retinas tends to fade away. Only when our attention is turned away from vision altogether, perhaps focused instead on our inner thoughts, do our eyes rest, and we are said to stare into space.

In addition to our physical control of the information reaching our brains through our sense organs, we can mentally focus our attention on a particular part of this information. For example, we are expert at picking out the voice of the person we are talking to from the hubbub of a party. We achieve this not only by recognising that person's voice from its pitch and volume, but also by following the words and even the meaning of the conversation. So focusing our attention on a particular conversation at a party involves processing a considerable amount of auditory information from other conversations, to the extent that we pick up the words and even something of the meaning of the other conversations. Our brains are quite capable of the low-level processing of so much auditory information, but we do not have the mental resources for the higher-level processing of so much information (nor, indeed, the mental and physical resources to respond to more than one conversation at a time). Focusing our attention on just one conversation, then, allows us to select the auditory information we consider most important for higher-level processing.

Our low-level processing of auditory information from other conversations also allows us to monitor them subconsciously. Even when someone is concentrating on one conversation at a party, she is immediately aware when her name is mentioned in another, because her brain has been subconsciously processing the auditory information from the other conversation at a low level. The level of our party-goer's ability to register the words and the meaning of the other conversation gives a good indication of the extent to which we process sensory information on which our attention is *not* focused. She can easily register the sound of a single word in another conversation, but she would have difficulty registering the meaning of a sentence. For example, if the other conversation includes the words 'that's Robert's girlfriend, the archaeologist', she will probably be alerted to the conversation by the mention of her boyfriend's name and of archaeology. But if the words were 'that's old Bob-boy's fancy chick', she may remain unaware that she is the subject of conversation. Our low-level processing of auditory information extends to understanding the words of other conversations but not always to understanding the meaning.

The intelligent computer, like us, would have limited mental resources for the high-level processing of information, and so would benefit from the ability to focus its attention. Attention is one of the many human attributes whose reproduction on a computer would make a fundamental contribution to its intelligence, rather than being an anthropocentric whim of the designer. It would be simple to provide an intelligent computer with the ability to control its sense organs. For example, the computer could be provided with motors to move the video cameras acting as its visual sensors in the same way that we humans move our eyes. It would also be relatively simple to provide the computer with the ability to allocate high-level processing resources to a particular part of the information from its sense organs. For example, once the computer has been provided with the ability to derive words from auditory information and the ability to derive meaning from words, it could be provided with the ability to select a particular conversation for high-level processing based on continuity of meaning (as well as simpler cues such as pitch and volume). The question that remains is how to provide the computer with the ability to decide *what* to focus its attention on. I will return to this important question in the chapter on emotions.

Memory

One of the most effective uses of computers is for the storage of information. Information is stored in a computer database in much the same way as it is written on forms. For example, a form for personal information has boxes labelled 'Title', 'Initials', 'Surname', 'Address', 'Telephone Number', 'Occupation', and so on, in which to write the relevant information (see Figure 10). The same rigid structure is imposed on a record of personal details in a computer database. The correct information must be stored in the relevant boxes, so that the computer can process the information according to the precise instructions of the programmer. If, for example, information about Mr A.G. Beech's occupation were stored in the address box rather than the occupation box, the computer would address letters to 'Mr A.G. Beech, Lawyer' instead of 'Mr A.G. Beech, 31 Meadow Road, Bridgeton, Lincolnshire'. This would make it difficult for the postman to know where to deliver the letters. Information is retrieved from a computer database

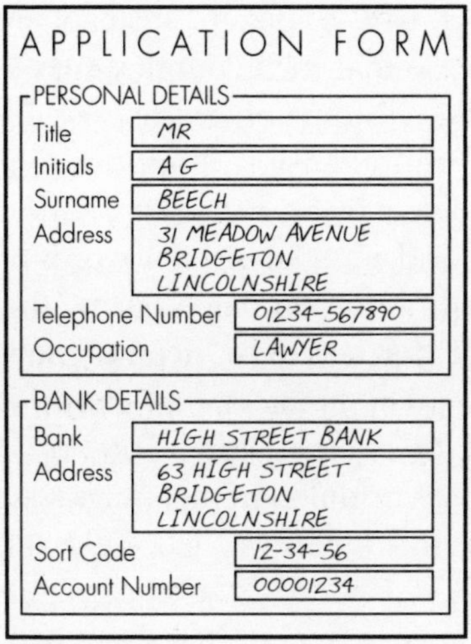

APPLICATION FORM

PERSONAL DETAILS

Title	MR
Initials	A G
Surname	BEECH
Address	31 MEADOW AVENUE BRIDGETON LINCOLNSHIRE
Telephone Number	01234-567890
Occupation	LAWYER

BANK DETAILS

Bank	HIGH STREET BANK
Address	63 HIGH STREET BRIDGETON LINCOLNSHIRE
Sort Code	12-34-56
Account Number	00001234

Figure 10. A form completed by Mr A.G. Beech. Information is stored in a computer database in much the same way as it is written on forms.

through the same rigid structure. For example, to retrieve the names of all the lawyers from the database, the computer would be instructed to retrieve all records with the word 'Lawyer' in the occupation box.

Human memory is considerably more flexible than a computer database. We remember different things about different people, such as the cheerfulness of the milkman, the long red hair of the girl next door, or the name of the boss's wife. We are reminded of people in countless different ways, so that a trip to Paris might remind us of someone we met when we were last there, or a particular turn of phrase might remind us of someone who habitually uses it, or the smell of a particular perfume might remind us of someone who wears it. The rigid structure of computer databases is thus inappropriate for the reproduction of human memory on a computer.

At its simplest, human memory is based on perception. For a short time after we hear someone speak, the sound of the speaker's voice remains in our memory. Such perception-based memory is essential if we are to derive meaning from what we perceive. When we listen to speech, we do not derive the meaning of each word as it is spoken. Instead, we listen to whole phrases at a time, so that such complex cues as the meaning of the words and the customs of grammar and idiom can contribute to our comprehension. Suppose a guard on a noisy railway station is asked by a passenger: 'Is this the right platform for trains to Cambridge?' The first few words of the question ('Is this the . . .') might be so drowned out by the noise of a train pulling into the platform as to be incomprehensible taken alone. The word 'right' taken alone is ambiguous. It could be interpreted as any of the words 'write', 'rite' or 'right', the last of which has a number of different meanings (right as the opposite of wrong, right as the opposite of left, right as in the passenger's rights, and so on). The noise of the train might again drown out the last half of the sentence, so that the guard can scarcely discern the words 'platform' and 'Cambridge'.

On the face of it, the guard might be expected to have great difficulty in understanding the passenger's question. In fact, he would understand it quite easily. The context in which he is asked the question, as a guard who works on the Cambridge line and is frequently asked about platforms, allows him to discern the words 'platform' and 'Cambridge'. If he hears the 'plat . . .' of 'platform', he need not wonder whether he is being asked about platinum or platypuses, because he knows that questions about platinum and platypuses are unlikely in the circumstances. Similarly, he need not distinguish the word 'Cambridge' from words such as sandwich or umbrage, only from other destinations such as Peterborough or York. Once he has understood 'platform' and 'Cambridge', the guard can derive the meaning of the rest of the sentence by drawing on his memory of the sound of the passenger's voice. Because he remembers the sound of the word 'right', rather than one of the many of the meanings of the word, he is now able to select the meaning that fits the surrounding words. Further, even though the words 'is this the' and 'for trains to' were too drowned out to recognise out of context, the cues of grammar, idiom and meaning now combine with the guard's memory of the sounds of these words to allow him to recognise them after the event.

We also use perception-based memory for mental manipulations. A driver might decide on his route by mentally picturing it either as he would see it on a map or as he would see it from the driving seat. A schoolgirl doing mental arithmetic might remember the intermediate results of her calculation by verbally rehearsing them in her mind. Psychologists' experiments suggest that the same perception-based memory is used whether the sights or sounds remembered are actual sights and sounds perceived by our eyes and ears, or imaginary sights and sounds produced by mental manipulations. Suppose that the passenger who asked about the right platform for trains to Cambridge then asks the guard about the return fare for one adult and one child. The guard uses the same perception-based memory for his verbal rehearsal of the fare calculation as he so recently used to remember the sound of the passenger's voice.

Our perception-based memory has a limited capacity. For example, the number of words it can hold seems to be fixed by how long it takes to rehearse the words. It is as if there is a loop of audio tape in the human brain that is long enough to record only a couple of seconds of sound. Unless we mentally rehearse the sound on the tape, the memory is rapidly lost. It makes sense that perception-based memories are used only for immediate mental manipulations and then lost in this way. If our permanent memories of sights and sounds were perception-based, all the benefits of processing sensory information would be lost when we memorised it. If we memorised extraneous as well as important sensory information, we would have to repeat our distillation of the important from the extraneous every time we recalled the information. If we memorised sensory information rather than the meaning we derived from it, we would have to repeat our derivation of meaning every time we recalled the information. Moreover, our brains would have to be enormous to retain unprocessed sensory information. So our permanent memories are based not on perception, but on meaning: we remember not sights and sounds, but the meaning we derive from those sights and sounds.

Psychologists have done much research aimed at discovering the nature of our permanent, meaning-based memory, and have formulated models of memory to match their findings. A certain class of models, called connectionist models, have recently gained popularity among psychologists. These models describe the higher-level processes of the human brain in terms of interconnected elements in an attempt to explain how the interconnected neurons in the human brain could give rise to these processes. A connectionist model of permanent, meaning-based memory, which I will refer to as the associative model, not only represents a successful model of memory in humans, but also provides the designer of the intelligent computer with a suitable model for the reproduction of memory on a computer.

Our permanent, meaning-based memories are encoded as entities. An entity may represent a category, such as the *category* of peaches, or a *specific* object, such as the peach I ate at lunchtime. An entity might represent a person, such as the milkman, a place, such as Shanghai, an action, such as jumping, or an abstraction, such as imagination. Entities are defined and described exclusively through associations with other entities (see Figure 11). The *category* of peaches might be associated with the category of fruits (of which it is an example), the colours red, orange and yellow, the soft, juicy flesh, the hard, pitted stone, the taste (which I am unable to describe), the smell (which I am again unable to describe), and so on. The category of peaches is

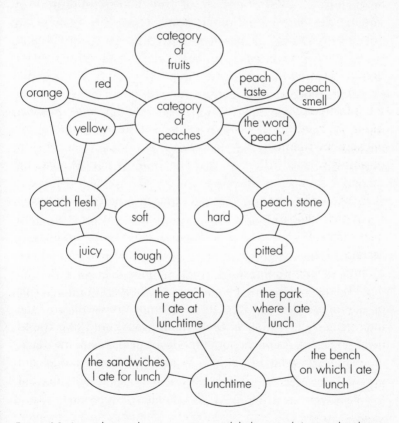

Figure 11. According to the associative model, the peach I ate at lunchtime is encoded in my permanent, meaning-based memory as an entity defined and described through associations with other entities.

also associated with the sound of the word 'peach', but this does not define the entity any more than the other associations. The *specific* peach I ate at lunchtime is associated with the general category of peaches, the general category of things I have eaten, various other events that occurred at lunchtime, the slight toughness of the flesh of the fruit, and so on.

When we recall something we have previously perceived, we reconstruct the perceptual details from the abstract representation in terms of entities and associations. So when I think of the peach I ate at lunchtime, I vividly recall the feel of sinking my teeth into the flesh, the taste of the juice running into my mouth, the sight of the pitted stone. This is not because my memories of eating the peach are perception-based, but because I am able to reconstruct the perceptual details of eating the peach from my meaning-based memories. The reconstruction seems vivid because it uses the same perception-based memory used for the original perception. The only difference between my experience of the reconstruction and my experience of the original perception is that the sensory information is missing, so the reconstruction is somewhat fleeting. Any details missing from my memory of the peach are supplied by my stereotype of a peach, so when I think of the pitted stone, although I see a vivid mental picture of it, the positions of the pits on the stone in my mental picture in no way correspond to the positions of the pits on the real stone. I did not convert this unimportant perceptual detail into meaning-based memory, so it is irretrievably lost. To obviate this loss, I conceive a random distribution of pits based on my stereotype of a peach stone, safe in the memory that the specific peach stone in question did not differ significantly from that stereotype. The fact that the exact details of the resulting mental picture may bear no precise relation to the original perception does not worry me, because I have no memory of the original to know that the reconstruction is inaccurate.

The associations for a specific instance of a category tend to represent the differences between the characteristics of the instance and the expected characteristics of the category. So the peach I ate at lunchtime is particularly associated with the slight toughness of its flesh, because usually peaches have softer flesh, but it is not directly associated with the colours red, orange and yellow (though it was indeed red, orange and yellow), because these are the usual colours for a peach. This device of remembering only the differences between a thing and its stereotype represents an important economy of memory.

One consequence of this device is that we tend to see the world as consisting of things that conform more or less to stereotypes. The general utility of this view can sometimes blind us to its limitations. We sometimes refuse to recognise that things can differ from their stereotypes, particularly when such differences are not immediately apparent. Racism provides an emotive example. Generalisations about people of a particular race may be more or less valid, but racism arises from the refusal to recognise that people of that race *can* differ from whatever stereotype we impose. The result is that our stereotype is not tempered by observation of individuals of the race in question, and so becomes distorted.

Another consequence of the device of remembering only the differences between a thing and its stereotype is that our brains are particularly alert to those differences that *are* immediately apparent, since it is the differences that must be remembered. Designers of advertisements are experts at exploiting this alertness to differences between expectation and reality. Slogans are printed in unusual lettering, trade names include words that are deliberately misspelt, photographs are retouched to give impossible images, all to attract the attention of brains sensitive to differences from the norm.

Memories are retrieved through the activation of entities. Activation spreads from entity to entity through associations. Consider a history student answering an examination question about taxation under Henry VIII. The entities representing taxation and Henry VIII are activated. The activation spreads to other associated entities, some relevant to the question, some irrelevant. Taxation is associated in the student's mind with various taxes, including relevant taxes from the period such as fifteenths and tenths, first fruits and tenths, land tax, goods tax and poll tax, along with irrelevant modern taxes such as income tax, sales tax, value added tax and capital gains tax. Henry VIII is associated in the student's mind with various people, including Cardinal Wolsey, Archbishop Cranmer, Thomas More and Thomas Cromwell, and various events, including marriages and divorces, invasions of France, the Reformation, the dissolution of the monasteries and several revolts. All of these entities are further associated with each other and with other entities in a complex web of associations (see Figure 12). Activation spreads through it so that the entities relevant to taxation under Henry VIII are highly activated. For example, the relevant taxes (fifteenths and tenths, first fruits and tenths, land tax, goods tax and poll tax), which are associated, however indirectly, with Henry VIII, are more highly activated than the irrelevant taxes (income tax, sales tax, value added tax and capital gains tax), which are associated only with taxation in general.

The stronger the associations that define and describe an entity, the easier it is for activation to spread to the entity, and so the easier it is to recall the memory it represents. Associations between entities are strengthened with use, so the history student revises her knowledge of Henry VIII to improve her chances of recalling it in the examination. Unused associations slowly weaken over time, so if the history student does not use her knowledge of Henry VIII after the examination, her ability to recall that knowledge will slowly diminish.

The more associations that define and describe an entity, the more paths along which activation can spread to the entity, and so the easier it is to recall the information it represents. The key to effective learning is the elaborate processing of the information to be learnt to form more associations between the entity representing that information and entities representing possible cues for its recall. This holds whether the learning is intentional, as for a student revising for exams, or incidental, as for a

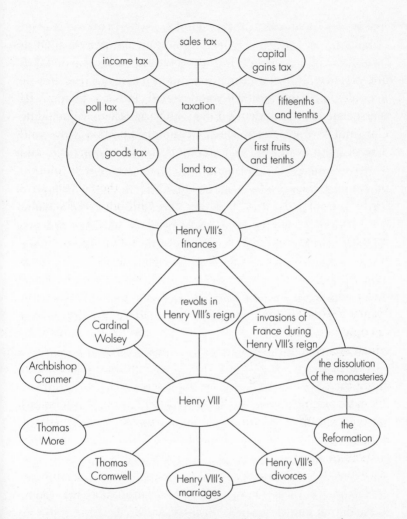

Figure 12. Memories are retrieved through the activation of entities. Activation spreads from entity to entity through associations.

secretary who has no particular intention of learning the contents of the letter she is typing. For the secretary, the elaborate processing might involve manually checking the spelling, grammar or content of the letter. Performing any of these processes creates more paths for activation, thus making her better able to recall the contents of the letter than if she had just typed it. For the student, elaborate processing of the material to be learnt is a far more effective study technique than simple rehearsal. In history, elaborate processing of a period to be learnt might involve exploring the motivations of the protagonists and the connections between the various events. In science, elaborate processing of a theory to be learnt might involve considering the consequences of the theory and applying it to solve problems. In less structured subjects such as languages, elaborate processing is more difficult. When learning vocabulary there are often no obvious associations to be made between a foreign word and its English translation, but there are alternatives to rote learning nonetheless. Learning the etymology of the foreign word may help by associating it with other words in the foreign language. Even nonsense associations have been found to aid learning. The French word '*poisson*', which means 'fish', might be learnt by associating it with the English word 'poison', and associating 'poison' and 'fish' by imagining a poisoned fish floating dead in a pond.

Association influences recall in powerful and sometimes unexpected ways. It has been shown, for example, that what we learn becomes associated with the surroundings in which we learn it, so that we can recall what we learn more easily in that environment than in any other. Hence the best place for a student to revise is in the examination hall.

We remember far more than we can recall. A common experience for a quiz enthusiast is to be unable to recall the answer to a general knowledge question, despite being quite sure that he knows it. Suppose that the contestant is asked the name of Henry VIII's first wife. He might recall the names of the other five of Henry VIII's wives (Anne Boleyn, Jane Seymour, Anne of Cleves, Catherine Howard and Catherine Parr), but be unable to recall the name of the first. If he were shown a list of possible answers (such as Margaret Beaufort, Elizabeth of York, Catherine of Aragon, Mary Tudor and Jane Grey), he would immediately select the correct one (Catherine of Aragon), so demonstrating that he does know the answer. He is unable to recall it without prompting because the associations between the entities representing cues in the question and the entities representing the answer are too weak, or are interfered with by other stronger, but irrelevant, associations. Henry VIII might be so strongly associated in his mind with Anne Boleyn that the activation of the entity representing Catherine of Aragon is insignificant in comparison.

I have drawn a distinction between entities and associations to allow the associative model to be explained in simple terms, but the model is improved if this distinction is abandoned. The association between the entities representing Henry VIII and Anne Boleyn can equally be regarded as an entity, one which might in turn be associated with the personal, political and religious circumstances of their relationship.

The formation of new associations between entities is prompted by coincidence of activation of those entities. Consider a child coming across an elephant for the first time. As the child sees the elephant, entities representing its various features are activated: entities representing the shapes of the subobjects into which the elephant is segmented, the pattern of the relative positions and proportions of these subobjects, the form and pattern of the elephant's facial features, the colour and texture of the elephant's skin, the sights and sounds and smells of the zoo in which the elephant is seen. If the child had seen an elephant before, the activation of all these entities would spread to the entity representing the category of elephants through the associations that define that entity. Instead, the child has not seen an elephant before, so no such entity exists. However, the coincidence of activation of all these entities prompts the formation of an association between them. Since associations and entities are one and the same, the newly formed association can equally be regarded as an entity. This newly formed entity represents the category of elephants.

This simple mechanism for the formation of associations is complicated by higher-level processes such as language comprehension. Consider someone overhearing the sentence 'Peter lives in London and Susan lives in Glasgow' at a party. If the formation of associations were prompted exclusively by coincidence of activation, the party-goer would form a strong association between London and Susan prompted by their being mentioned in quick succession. Instead, the process of language comprehension intervenes to ensure that the party-goer correctly associates Peter with London and Susan with Glasgow. Nonetheless, the party-goer might form some association between Peter and Susan, even though no such association is made explicit in the sentence. This association might be meaningless, but it might equally represent some meaningful relationship (Peter and Susan might have been mentioned in the same sentence because they are brother and sister).

Generalisation is inherent in the associative model of memory. Consider a child forming an entity to represent the next-door neighbour's dog Brutus. This entity would initially be crudely defined. It would be applied to all small, four-legged animals, not just dogs, and not just Brutus. At this stage, then, the child might use the word 'Brutus' to refer to all dogs and cats, much to her parents' amusement. Further experience of small, four-legged animals might prompt the child to recognise that some are large and have pointed noses, and some are smaller and have flatter noses, and so form entities representing the distinct categories of dogs and cats. Now the child might use the word 'Brutus' to refer to all dogs, but learn to use the word 'cat' to refer to cats. Finally, the child would recognise that the instances of dogs that she frequently sees in the next garden are the same individual, whereas the various instances she sees in the street are not. She would then form a separate entity representing Brutus, associating the word 'Brutus' with that entity, and the word 'dog' with the existing broader entity representing the category of dogs. The child's concept of dogs is derived not from the examination of individual dogs and the recognition of the similarities between them, but from the refinement of a cruder category formed from her first experience of a small, four-legged animal. It is the process of refining categories, rather than that of generalising instances, that demands the attention of the designer of the intelligent computer.

Prior knowledge has a significant influence on the encoding of new knowledge. Because a new entity is defined through associations with existing entities, the number and appropriateness of the existing entities affects how well the new entity is defined. For example, if a rambler with little knowledge of archaeology were to discover a prehistoric stone arrowhead, he might define it as a small, pointed stone, without even recognising it as an arrowhead. An archaeologist, on the other hand, discovering the same arrowhead, might define it in terms of what stone it is made of, what tools would have been used to make it, what animals it would have been used to kill, how it would have been fixed to the arrow, how well finished it is, and so on, as well as by its precise shape and size. The archaeologist's prior knowledge of arrowheads provides a large number of entities with which the specific arrowhead can be appropriately associated, so she remembers far more about it than the rambler. Existing entities influence not only the comprehensiveness with which the new entity is defined, but also the emphasis with which it is defined. If the archaeologist is interested in how prehistoric humans manufactured arrowheads, she will define the arrowhead more in terms of the evidence for its method of manufacture, but if she specialises in the population distribution of prehistoric humans, she will define it more in terms of its position with respect to known prehistoric settlements.

The influence of prior knowledge on the encoding of new knowledge extends beyond how we remember things to how we perceive things. What we perceive is not sensory information, such as the photons falling on our retina, but the meaning we derive from it, the trees and fields and sky resolved from those photon-falls. This meaning does not merely *influence* our perception, it *is* our perception. Our prior knowledge, through its influence on how we encode this meaning, has a significant influence on how we perceive things. Faced with what adults would call a cat, the child who has not yet learnt to differentiate between dogs and cats would see not a cat but a small four-legged animal. The rambler discovering the arrowhead not only remembers less about it than the archaeologist, he perceives less; the archaeologist covering the same ground as the rambler would be more likely to notice the arrowhead in the first place. An Englishman with little previous contact with Chinese people, watching a crowd at a Chinese New Year celebration, would see little difference between each member of the crowd, whereas a Chinese person would distinguish each individual as quite different from the next. The failure of the child, the rambler and the Englishman to notice the same details as the adult, the archaeologist and the Chinese person is not a result of the failure of the former to look closely enough. It arises from real differences in the way they encode meaning, real differences in perception.

If human memory involves perception-based encoding for immediate processing of sensory information and meaning-based encoding based on entities and associations for permanent memories, how could perception-based and meaning-based memory be reproduced on a computer?

The same perception-based memory is used for the perception of sensory information and the manipulation of mental images, but perception-based memory is different for each of the different senses. Visual information is encoded at the lowest levels as spots, lines, edges, textures and colours, and at higher levels as simple geometrical shapes representing objects and subobjects. Auditory information is encoded at the lowest levels as intensities of different pitches of sound, and at higher levels as phonemes, words, intonations and tones of voice. The encoding of sensory information in these various different ways is not incidental to perception but fundamental to it, and the processes of perception could not be reproduced on a computer without also reproducing these encodings. Once they have been reproduced on a computer, the reproduction of perception-based memory would be simple. The designer of the intelligent computer need only take the perception-based memory used to encode sensory information for the processes of perception, and allow it to be also used for the manipulation of mental images. Nor would the formation of such mental images be difficult to reproduce, since it is the reverse of the processes of perception. While perception derives meaning-based memories from perceptual details, mental images are formed by the reconstruction of perceptual details from meaning-based memories.

The reproduction of meaning-based memory on a computer would also be reasonably simple. The associative model forms a descriptive model of meaning-based memory in humans, but it also provides a prescriptive model for the reproduction of meaning-based memory on a computer. For the designer of the intelligent computer, it does not matter whether or not future research reveals the associative model to be an accurate model of human memory, whether or not the interconnections of entities in the associative model turn out to correspond to the interconnections of neurons in the human brain. What matters is that a computer with a memory based on the associative model would have the same memory-related characteristics as humans. It would not be difficult to provide the computer with a precise set of instructions to reproduce the strengthening of associations through use, the weakening of associations over time, the spreading of activation from entity to entity through associations, the formation of associations prompted by coincidence of activation, and so on. As discussed, the formation of associations is complicated by higher-level processes such as visual perception and language comprehension. But again, once these processes have been reproduced on a computer, the provision of instructions for the formation of new associations would be reasonably simple.

The most challenging difficulties the designer of the intelligent computer would face in replicating human memory would be practical ones. Providing a computer with a memory as large as that of a human, and making the computer fast enough that it can manage such an enormous memory, would present real difficulties.

Perception II

The questions left unanswered at the end of the earlier discussion of perception can now be readdressed in the light of the associative model of memory. One of these concerned how a computer could be made to learn correlations between the different cues for distance determination and scene segmentation.

Consider distance determination. As discussed earlier, computers could be programmed to detect cues for distance determination, such as the speed at which features move across the computer's field of vision and the disparity in the positions of features in the computer's left and right fields of vision. The computer could be further programmed to activate different entities according to its observation of these cues, activating one particular entity if it detects a feature moving slowly across its field of vision, and another if it detects a feature moving quickly. Similarly, it would activate one particular entity if it detects a small disparity in the positions of a feature in its left and right fields of vision, and another if it detects a large disparity.

Whenever the computer sees near objects, features move quickly across its field of vision, and there is a large disparity in their positions in its left and right fields of vision, so the entities representing quick movement and large disparity would be activated by near objects. Conversely, whenever the computer sees distant objects, features move slowly across its field of vision, and there is only a small disparity in their positions in its left and right fields of vision, so the entities representing slow movement and small disparity would be activated by distant objects. The computer would observe object after object conforming to this pattern and, according to the associative model of memory, the invariable coincidence of activation of the entities representing quick movement and large disparity would prompt the formation of an association between these two entities. Similarly, the invariable coincidence of activation of the entities representing slow movement and small disparity would prompt the formation of an association between *these* two entities. Since associations and entities are one and the same, these newly formed associations could equally be regarded as entities representing short distance and long distance respectively. Their formation constitutes the computer's learning of correlations between the cues for distance determination.

The combination of the cues for distance determination follows from their correlation. Whenever the computer sees a near object, the activation of the entities representing quick movement and large disparity would spread to the entity representing short distance, and the computer would recognise the object as near. Similarly, whenever the computer sees a distant object, the activation of the entities representing slow movement and small disparity would spread to the entity representing long distance, and the computer would recognise the object as distant. Other cues for distance determination, including contextual cues, would contribute to the activation of the entities representing short distance and long distance through further associations (see Figure 13 on page 58).

The combination of the cues for distance determination in this way is extremely flexible, allowing for the determination of the distance of an object based on only one or two cues, if these are all that are available. For example, if the one or two cues available indicated long distance, then the entity representing long distance would be activated less strongly than if more cues were available, but it would still be activated more strongly than the entity representing short distance. The combination of cues in this way also allows for ambiguity, and for the resolution of ambiguity. For example, if those cues that require no prior knowledge of the objects were ambiguous, the entities representing short distance and long distance would initially be activated equally. However, once the computer had arrived at a preliminary assessment of the objects, those cues that do require prior knowledge of the objects could contribute to distance determination. They would favour the activation of one or other of the entities representing short distance or long distance, and so resolve the ambiguity.

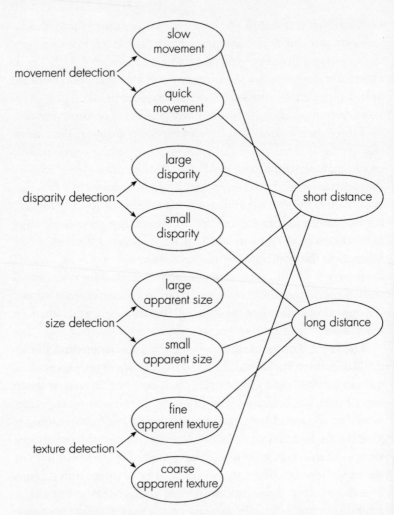

Figure 13. Entities representing short distance and long distance are formed as associations between entities activated according to the detection of the cues for distance determination.

So the associative model, previously discussed as a model for the reproduction of meaning-based memory on a computer, also allows the computer to learn correlations between the different cues for distance determination and scene segmentation, and to combine those cues in a flexible way. The distinction between the learning of facts, such as the fact that peaches have soft flesh and the fact that Peter lives in London, and the learning of concepts, such as the concept of distance and the concept of objects, can now be seen to be artificial. Facts and concepts alike are learnt through the formation of associations between entities prompted by coincidence of activation of those entities.

The other questions left unanswered at the end of the earlier discussion of perception concerned how a computer could be made to learn repertoires of simple geometrical shapes and common spatial patterns, and how these shapes and patterns appear in different orientations. It would be simple to program a computer to detect two-dimensional shapes such as circles and rectangles, and two-dimensional patterns such as the formations of the dots on dice. The computer could be further programmed to activate different entities according to its detection of these two-dimensional shapes and patterns. For example, it could be programmed to activate one particular entity if it detects a circle, and another if it detects a rectangle.

The key to learning repertoires of three-dimensional shapes and patterns is learning how the three-dimensional shapes and patterns appear as two-dimensional shapes and patterns when seen in different orientations. A cylinder appears as a circle when seen end-on, as a rectangle when seen side-on, and as a combination of a circle and a rectangle when seen in other orientations (see Figure 14 on page 60). By observing changes in the appearance of cylindrical objects or subobjects as they move, the computer would learn that these different combinations of circles and rectangles represent the same three-dimensional shape. When the computer sees a cylindrical object or subobject move from an end-on orientation to a side-on orientation, its appearance would change from circular to rectangular, so that the entity representing circles and the entity representing rectangles would be activated in quick succession. The coincidence of activation of these different entities would prompt the formation of an association between them. Once again, since associations and entities are one and the same, this

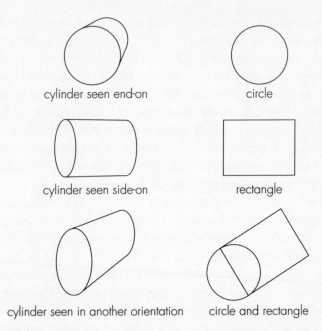

cylinder seen end-on

circle

cylinder seen side-on

rectangle

cylinder seen in another orientation   circle and rectangle

Figure 14. A cylinder appears as a circle when seen end-on, as a rectangle when seen side-on, and as a combination of a circle and a rectangle when seen in other orientations.

newly formed association could equally be regarded as an entity representing a cylinder. The formation of this entity constitutes the computer's learning of how cylinders appear in different orientations.

Having formed this new entity, the computer would be able to interpret objects and subobjects that appear as circles or rectangles as cylinders, and so add the cylinder to its repertoire of simple geometrical shapes. Of course, the interpretation of circles and rectangles as cylinders cannot be made with absolute certainty, since a circle may equally be interpreted as a sphere, and a rectangle may equally be interpreted as a rectangular block. Again, the associative model is flexible enough to handle such ambiguity. When the computer sees a circle, the activation of the entity representing circles would spread both to the cylinder entity and to the sphere entity. The ambiguity would be resolved, though, through further contributions to the activation of the cylinder and sphere entities, such as contributions from colour, texture and context.

When discussed in this way in terms of the associative model, object recognition is revealed to be no more than the process of encoding visual information. Objects are encoded in terms of associations between entities representing three-dimensional shapes and patterns. If the computer sees an object it has never seen before, the image of the object would prompt the activation of entities representing three-dimensional shapes and patterns in the computer's repertoire that would never before have been activated simultaneously. The coincidence of activation of these entities would prompt the formation of an association between them that could equally be regarded as an entity representing the geometry of the unfamiliar object. If, instead, the computer sees an object that it *has* seen before, the image of the object would prompt the activation of the same entities representing three-dimensional shapes and patterns as when the object was seen previously. This activation would spread to the entity previously formed to represent the geometry of the object, and so contribute to the computer's recognition of the object.

The above account of the processes of distance determination, scene segmentation and object recognition is much simplified. With distance determination, we do not distinguish only between short distance and long distance, but between the complete range of distances, from too close for comfort to the far horizon. We do not determine the distance of a single set of features at a time, but simultaneously determine the distances of all the objects we see. If these complexities are to be reproduced on a computer, the model of distance determination presented above must be realised on a much grander scale than previously suggested. Whatever the scale of realisation, though, the models I have presented provide a solid foundation for the reproduction of the processes of visual perception on a computer.

Simple feature detection, distance determination, scene segmentation and object recognition are not the only processes of visual perception; nor, of course, is vision our only sense. However, these processes are representative of the range and complexity of the processes of human perception. Indeed, the enormous amount of information our eyes collect and the abundance of knowledge we derive from it make visual perception the most difficult to reproduce on a computer. Perception of movement, contact, pressure, temperature, taste and smell would be easy to reproduce in comparison.

Skills

Humans manipulate and apply knowledge to a variety of different ends. We have motor skills, such as typing, driving a car, playing football and playing a piano, for the physical manipulation of our bodies and our environment. We have social skills, such as the comprehension and generation of spoken language, for interaction with other humans. We have mental skills, such as spelling, calculating, solving problems and making decisions, for the manipulation of information. Many activities require skills in more than one of these categories. Driving a car requires motor skills for manipulating the controls, social skills for interacting with other drivers, and mental skills for deciding how to react to different circumstances. All these ends – motor, social and mental – are quite different, but the means by which we achieve them are similar. Whether we are manipulating our limbs, our associates or our thoughts, our skills involve the manipulation of information, either information derived immediately from our senses or information recalled from our memories, according to learnt procedures.

A newborn baby appears to have no skills. His muscles are sufficiently well developed that he can move his limbs, but he lacks the motor skills to coordinate those movements. Both muscles and motor skills develop slowly as the infant learns to reach, grasp, crawl and walk, and he will continue to acquire further motor skills throughout childhood and adulthood. The newborn baby also appears to lack social and mental skills, and these, too, are acquired throughout childhood and adulthood. The obvious conclusion that all our skills are learnt should be approached with caution. The lack of evidence for skills in newborn babies may indicate only that psychologists are as yet unable to detect such skills, rather than that they do not exist. Further, just as the development of our muscles after birth is determined by our genes (albeit influenced by environmental factors), so might the development of skills after birth be determined by our genes rather than by our experience, and so be innate rather than learnt. Nonetheless, children are able to learn a variety of different skills, depending on their environment. A child born into a nomadic African tribe may learn to speak a language of clicks and tones and to collect plant matter for food, whereas a child born into a Western European setting may learn to speak a language of consonants and vowels and to handle a knife and fork. Such differences, along with the fact that we continue to acquire new and diverse skills throughout adulthood, suggest that our skills are indeed largely learnt.

Skills may be learnt by investigation, imitation or instruction. Perhaps the most significant example of learning by investigation is that of the infant learning to coordinate his movements. Later on other skills are learnt by investigation too, as when a child who has reached the limits of the programming manual supplied with his computer carries on to learn further programming skills by trial and error, or when a child learns ball skills by practising with a football on his own. But most of what we learn is from others, by imitation or instruction. We learn to drive a car through a combination of watching other people drive and taking lessons with an instructor. We learn to generate and understand language, and interact socially in other ways, largely through imitation of our associates and role models (instruction has been found to contribute remarkably little to a child's learning of language, and I doubt that people still use books of etiquette to learn social skills). We learn spelling and arithmetic from teachers at school through a combination of instruction, whereby the teacher presents the rules to be followed, and imitation, whereby the teacher presents examples of the application of those rules.

Motor skills are among the earliest we learn. Our muscles are controlled from our brains through the transmission of signals along motor neurons, whose activation causes our muscles to contract. Each motor neuron is connected to a number of other neurons in the brain, and if enough are activated, then the motor neuron, too, is activated, causing the muscle to which it is attached to contract. A newborn baby's uncontrolled waving of his limbs, then, is the result of the random activation of motor neurons causing the random contraction of the muscles in his limbs. This suggests that the activation of the neurons in the newborn's brain to which the motor neurons are connected is also random.

The crux of the coordination of the movements of our limbs is the correlation of the activation of motor neurons and the activation of various sensory neurons. An adult would interpret these correlations as being between her moving her arm and the various sensations that result from this movement, such as the sensations of muscles moving under her skin, of air flowing across her skin, of contact between her arm and an object, and of the image of her arm moving across her field of vision. Whenever she moves her arm, these are the sensations that result. In contrast, the newborn baby does not yet understand the world in terms of his controlling the motion of his limbs through space. For him, the unconscious correlations between the activation of the motor neurons connected to the muscles in his limbs and the activation of the sensory neurons connected to the receptors in his skin and eyes are abstract. These correlations will eventually form the basis of his concepts of motion and space, and even his concepts of self and will.

Amateur footballers tend to be particularly aware of the difference between knowing that kicking the ball in a certain way will send it hurtling into the back of the net, and being able to coordinate their movements to achieve this. Improving our motor skills involves reinforcing and refining our knowledge of the correlations between our movements and their perceived effects. By repeatedly attempting different movements and observing their effects, we develop our ability to coordinate our movements to achieve the effect we want. In other words, practice makes perfect. This is not to say that we cannot learn motor skills from others. A football coach, with his wide knowledge of the correlations between movement and effect pertinent to ball skills, can focus a player on the particular aspects of his movements that might achieve the desired effect, and suggest how those movements might be improved. But the player can only benefit from the coach's insights if he is able to learn how to coordinate his movements in the revised way the coach has suggested. This can be achieved only through practice.

The nimble leap from examination of learning the correlations of motor coordination to that of coordinating our movements avoids two important questions. The first is the question of motivation. What induces us to move our muscles in ever more refined ways, and so to reinforce and refine the correlations we learnt from our first, random movements? The second is the question of will. What is the nature of our decisions to move or not to move our muscles, and to move them in one particular way rather than another? Our motivation and our will are fundamental to our humanity, and are central to any discussion of whether computers can be like humans, and are examined in detail in the chapters on emotion and consciousness.

Questions of motivation and will aside, the learning of motor skills could be reproduced on a computer through the associative model of memory. The computer would be provided with a robotic body with motors controlled through the activation of motor entities, which would be activated and associated in the same way as other entities. The difference is that the activation of a motor entity would prompt the operation of a motor in the robotic body. Just as distance determination and scene segmentation involve learning correlations between the activation of the entities representing various cues, so motor coordination involves learning correlations between the activation of motor entities and the activation of entities representing the resulting sensations. Whenever the computer activates a motor entity, a motor in the robotic body would operate, the robotic body would move, and entities representing the resulting sensations would be activated. For example, whenever the computer operates a motor in its robotic arm, entities representing the sensations that result from the movement of the arm, such as the tactile sensation of contact between the arm and an object and the visual sensation of the arm moving across the computer's field of vision, would be activated. The invariable coincidence of activation of the motor entity and the entities representing these sensations would prompt the formation of an association that constitutes the computer's learning of motor coordination.

Initially, the tactile and visual sensations resulting from the computer's movement of its robotic arm would be abstract, but as the computer develops concepts of distance and objects, it would interpret these sensations in terms of the movement of its robotic arm, and so form concepts of motion and space. Having learnt that the activation of particular motor entities prompts particular movements of its robotic body, the computer would also form concepts of self and will. Its concept of self would encompass those parts of the environment it can control through the activation of motor entities, and its concept of will would encompass the phenomenon whereby the activation of motor entities prompts the movement of particular parts of its self. (Since the computer would learn mental as well as motor skills, its concepts of self and will would encompass mind as well as body.) In addition, having learnt the correlations of motor coordination, the computer would proceed to learn motor skills, encoded as series of activations of motor entities.

Mental skills involve the manipulation of information according to a procedure. For example, a child might be taught to add two single-digit numbers according to the following procedure.

To add five and three, she should count five counters into a pile, then count three more counters into the pile, then count the total number of counters. The child's successful completion of an addition according to this method requires that she knows how to count and that she understands the procedure for addition. It turns out that these two skills are learnt in quite similar ways.

Counting three or five or eight counters can be achieved by arranging the counters in a line, then reciting the numbers one, two, three, and so on, pointing to each of the counters in turn. The number recited when the last counter is reached indicates the number of counters. The child learns to recite the numbers from one to ten in much the same way as she learns the letters of the alphabet or the words of a nursery rhyme; indeed, when she is learning the numbers or the letters, she may be encouraged to fit them to a rhyme to aid her learning. Whether it is numbers, letters or nursery rhymes, what the child is learning is a series of words (in the case of the letters of the alphabet, words that represent the sounds of the letters).

Psychologists have done much research on our ability to remember series of items and have a good understanding of how series are encoded in human memory. Their work provides the designer of an intelligent computer with the insights required to adapt the associative model of memory to encompass the encoding of series. Like other permanent, meaning-based memories, the items in a series, along with the series itself, are encoded as entities, but the associations that define the series are different from the simple associations previously discussed. One difference is that there is always a strong association between the entity representing the series and the entity representing the first item in the series, and a weaker association with the entity representing the last item. This means that the first and last items can be easily recalled, but the intermediate items are generally recalled by running through the series from the beginning. Of course, this would be cumbersome if the series were long, but this problem is avoided by splitting long series into more manageable chunks, to give a series of subseries. One of the advantages of fitting numbers or letters to a rhyme is that the phrases or lines of the rhyme serve to define suitable subseries of numbers or letters (see Figure 15). Another difference is that the associations that allow us to run from one item in the series to the next do not work so well in reverse. It is more difficult to say the alphabet backwards than to say it forwards. The formation of new associations between entities representing items in a series is prompted by coincidence of activation of the entities, in turn prompted by mention of the items in quick succession.

Once she has learnt to count, the child will learn the procedure for addition as a series of words describing the tasks to be performed: 'Count out five. Count out three. Count them up.' This series of words will be learnt in the same way as all the other series of words mentioned so far – the numbers from one to ten, the letters of the alphabet, the words of a nursery rhyme. Again, the child may be encouraged to fit the words describing the procedure to a rhyme to allow her to remember them more easily. She will initially speak the words as she performs the

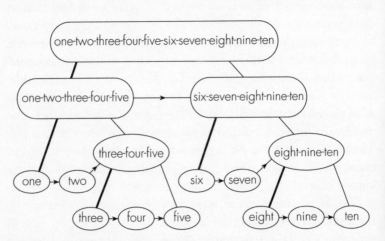

Figure 15. Entities representing the series of numbers between one and ten learnt according to the rhyme: 'One, two, three-four-five, once I caught a fish alive. Six, seven, eight-nine-ten, then I threw it back again.'

tasks, and by following the rules as she speaks them, she will successfully complete the addition. Since the tasks of the addition procedure are novel for the child, entities will be formed to represent the tasks. As she practises the procedure, these entities will be repeatedly activated in succession, leading to the formation of a new entity representing the series of tasks prompted by coincidence of activation of the other entities. Eventually, the child will no longer recall the rules for these tasks as a series of words but automatically recall the series of tasks themselves. She has learnt the addition procedure.

Adding two numbers using piles of counters does not appear to be a mental skill, as it is essentially linked to the physical world, but the child will soon learn to manipulate numbers in her head in the same way that she manipulates the counters. The next addition procedure she learns may involve counting upwards from five in her head, extending one finger every time she counts one in her head, until she has extended three fingers. Eventually she will have performed the addition of five and three so often that she will have associated the entities representing the numbers five and three with the entity representing their sum. At this stage she will be able to recall the sum of eight immediately, without recourse to an addition procedure. She will then be able to use this knowledge in more advanced procedures. For example, the child's successful completion of the addition of 125 and 753 according to procedures for addition commonly taught in schools requires that she know how to add any two numbers from nought to nine and that she know the procedure for addition of numbers of several digits. Throughout her life, she will learn more and more advanced mental skills based on the simpler mental skills she already knows.

This progression, from learning the rules for a procedure, through learning the tasks of the procedure, to learning the procedure to the extent that its execution becomes automatic, is common to all mental skills. If a mental skill is learnt by imitation rather than instruction, the first stage of learning explicit rules for the procedure is skipped, and the tasks of the procedure are learnt more directly. Psychologists have found that we learn best if we are presented both with the rules for a procedure, so that we learn by instruction, and with examples of the application of those rules, so that we learn by imitation. Neither instruction nor imitation on its own is so effective.

To describe our learning of mental skills in terms so mundane as procedures and tasks may seem to derogate human intelligence. But just as a meteorologist's understanding of the atmosphere in terms of fronts and pressure systems need not reduce his appreciation of a beautiful sunset, and a historian's understanding of the causes of a war in terms of politics and economics need not reduce her horror of the war, so our understanding of mental skills in terms of procedures and tasks need not reduce our reverence for human intelligence. Further, it may seem unlikely that the achievements of humanity's greatest scientists can be described in such mundane terms. There is indeed more to the achievements of scientists than the acquisition and application of knowledge and mental skills, for creativity is also required. Nonetheless, the mental skills employed by scientists during most of their work are not fundamentally different from the mental skills employed by a child to perform mental arithmetic, and the same is true of mathematicians, philosophers, historians and sociologists. Expertise in any of these fields requires considerable knowledge, considerable mental skills, and, for a master in the field, a touch of creativity. To understand human learning of knowledge and mental skills, to understand human creativity, is not to disparage human achievement.

This discussion of the learning of skills has examined motor skills and mental skills, but not social skills. In particular, it has not covered the comprehension and generation of language. It is often said that language is fundamental to thought. The origin of this misconception is probably people's awareness of forming their clearest thoughts into words and sentences, recited in their minds. If someone is thinking about his weekly shopping, he might recite in his mind: 'I must remember to buy more milk this week.' If, instead, he is thinking about pouring milk into his tea and notices that not much milk is left in the fridge, he might still recite the word 'milk' in his mind, but might not form his thoughts about buying more milk into so coherent a sentence as before. This suggests that the recitation of words and sentences in the mind is not fundamental to thought, but secondary to it. Thoughts involve the manipulation of abstract information, and the formation of thoughts into words and sentences follows as an expression of this manipulation. The human brain is specially adapted to language comprehension and generation, and specific parts of the brain are dedicated to these functions. If the intelligent computer is to be as adept at language comprehension and generation as humans, it too must be specially adapted to these functions, and research is required to determine how these adaptations could be reproduced on a computer. However, regardless of such special adaptations, and regardless of the importance of language for communication, language comprehension and generation are ultimately no more than learnt skills, no more fundamental to intelligence than motor skills or mental skills.

The associative model, then, originally discussed as a model for the reproduction of meaning-based memory on a computer, can achieve much more than this. It allows the computer to learn correlations: between the different cues for distance determination and scene segmentation to achieve visual perception; and between its moving its robotic body and the various sensations that result from the movement to achieve motor coordination. The associative model allows the computer to learn skills. One of the developments of the model described in this section allows entities to represent not only information derived from the computer's senses, but also information destined for the computer's motors. For example, an entity may represent a task such as a physical manipulation of counters, perhaps as one of a series of tasks in a procedure for addition using counters. A further development allows entities to represent information not directly related either to the computer's senses or to its motors. For example, an entity may represent a task such as a mental manipulation of figures, perhaps as one of a series of tasks in a procedure for mental addition. The step from the representation of things external to the computer, either perceived through senses or to be manipulated through motors, to the representation of things internal, is important. It is an indication that the computer can know its own mind.

Logic and Science

Understanding of intelligence is often clouded by a misconception. Logic and science are afforded a far higher status in our society than perhaps they deserve. One of the most effective ways to discredit an adversary in an argument is to accuse him of being illogical, and one of the most effective ways to discredit a world-view is to call it unscientific. Further, people perceived to be highly logical and scientific thinkers, such as Albert Einstein, are exalted as the epitome of genius. Such exaggerated respect for logic and science gives rise to the misconception that intelligent computers must think logically and scientifically.

Logic is the application of a number of simple rules to statements known as premises to generate further statements known as conclusions. The rules of logic are such that if the premises are true, and the rules are applied correctly, then the conclusion is necessarily true. For example, from the premises 'Rudolf is a reindeer' and 'all reindeer have antlers', the conclusion 'Rudolf has antlers' can correctly be deduced.

One problem with logic is that the truth of any conclusion depends on the truth of the premises. For example, because reindeer periodically shed their antlers, the premise 'all reindeer have antlers' is not quite accurate. So the conclusion 'Rudolf has antlers' may not be true (Rudolf may have just shed his antlers). The problem cannot be solved simply by ensuring that the premises are true, because true premises are difficult to come by. There is no logical way of proving that a premise is true, other than by deducing it from other premises – which leaves the problem of how to prove these other premises.

Another problem with logic is that its very precision renders it inappropriate for expressing the imprecise ideas involved in human thought. For example, the statements 'all animals have legs' and 'all fish are animals' seem reasonable statements when taken separately, but taken together lead to the conclusion that 'all fish have legs'. The reason for such a nonsensical conclusion is that the word 'animal' is ambiguous. Sometimes it refers just to mammals, and sometimes not only to mammals but also to reptiles, amphibians, fish, insects, and so on. The problem is not peculiar to this example. Most human ideas defy translation into a precise logical form. The imprecision of human ideas is not necessarily a shortcoming. After all, we are capable of flexible thought in spite of, or perhaps because of, this imprecision. It is just that logic is inappropriate for the expression of these ideas.

Considering the fundamental limitations of logic and its inappropriateness for expressing human ideas, it is unsurprising that the spread of activation between associated entities, rather than logic, provides the basis for human thought. Consider a pre-school child's reaction to the statement 'Rudolf is a reindeer'. If she has any previous experience of reindeer, perhaps from pictures in books, the mention of Rudolf being a reindeer will conjure up an image of Rudolf as a large four-legged animal with antlers. The idea that all reindeer have antlers is understood by the child through her association of reindeer with antlers, rather than by any formal rule, and the idea that Rudolf has antlers is an assumption based on this association, rather than a deduction based on the rules of logic. Consider the reaction of the same child to the statements 'endocrine glands secrete hormones' and 'the thyroid is an endocrine gland'. Suppose she hears these statements as the same pre-school child who so successfully decided that Rudolf has antlers, at an age when she has yet to learn the rules of logic. Her reaction to the statements containing so many unknown words will be one of bemusement.

Suppose, instead, that she hears these statements in a biology lesson many years later. By now she will be thoroughly versed in the rules of logic, though perhaps unable to state them explicitly, so, even though she has never heard of the thyroid or come across the idea that hormones are secreted by glands, she can nonetheless conclude that 'the thyroid secretes hormones'. Logic is a mental skill for the manipulation of information, learnt in the same way as any other.

If logic is merely a mental skill rather than the basis of human thought, it is nonetheless a powerful tool. It can be used to discuss ideas, to demonstrate the consistency of one's own world-views and the contradictions of others', or vice versa. It can be used to consider the possible consequences of one's actions in advance. It allows precision of thought not otherwise possible, and so is invaluable in all academic disciplines and many practical pursuits.

Like other powerful tools, logic is dangerous when misused. It can be used inappropriately or inaccurately to justify or rationalise an unpalatable opinion or action, and to pretend that it represents the culmination of a logical argument rather than the product of basic emotions. This is particularly prevalent in our society, which holds logic to be superior to emotions. Such rationalisations are unhelpful or dangerous because they lend a patina of respectability to the opinion or action while masking the true emotions that gave rise to it. A couple may indulge in endless arguments over whether the other is acting reasonably or unreasonably without ever admitting the true reasons for their hostility. A proponent of the death penalty may use a battery of logical arguments to hide his simple emotional response that murderers should be killed, while his opponent may use a similar battery of logical arguments to hide his own simple emotional response that even murderers should *not* be killed. The politician who constructs an historical justification for conflict with a neighbouring country is supplying his compatriots with a rationalisation to mask the nationalist emotions he is inciting.

Just as logic does not provide the basis for human thought, nor should it provide the basis for computer thought. Nonetheless some programmers interested in reproducing intelligence on computers have found the lure of logic irresistible, perhaps because it is so simple to encode the rules of logic on a computer.

In one such logic-based approach, a research team programmed a computer to perform logical analysis of sentences in the English language. The computer was fed thousands of statements of general knowledge. It then pointed out contradictions between the statements and derived conclusions from them. The experiment was a fascinating exercise in logic, but the computer was not intelligent. One possible objection to the claim that the computer was intelligent is that it had no understanding of the information it was manipulating, in that it could not relate the term 'reindeer' to an animal which it could sense and interact with. This argument lacks force. The computer should not be condemned as unintelligent merely because it was immersed in a different environment from us (albeit a strange environment of general knowledge statements) and had a different set of sensors and motors. (The computer had just one sensor, the keyboard used by the researchers to type the statements, and just one motor, the screen on which its conclusions were displayed. The motor was an impotent one, allowing the computer to communicate with the researchers, but not to manipulate the environment.

There are, however, more serious objections to the claim that the computer was intelligent. The researchers programmed the computer to understand the rules of logic and the grammar and idioms of the English language, so any intelligence required to manipulate the statements of general knowledge was that of the researchers, not the computer. Moreover, at the end of each day of feeding statements into the computer, the researchers analysed its conclusions, many of which were contrary to common sense. They provided the computer with counter-statements to correct the error that led to each of these nonsense conclusions, so the intelligence required to distinguish sense from nonsense was again that of the researchers, not that of the computer. The computer was merely a tool for analysing sentences in the English language according to fixed rules of grammar, manipulating information derived from those sentences according to fixed rules of logic, and recording the results. It was no more intelligent than a pocket calculator with a memory.

Just as humans learn logic as a mental skill for the manipulation of information, so should computers. Again, this is an anathema to some programmers, so enticed by the simplicity of encoding the rules of logic on a computer that they find it perverse to leave the computer to learn logic for itself. But intelligence cannot be programmed, only the appearance of intelligence.

Scientific thinking is no more the pinnacle of human intelligence than logical thinking. Scientific knowledge and mental skills are learnt in schools and universities in the same way as knowledge and mental skills in other subjects. Science differs from other subjects in that there is more emphasis on mental skills, such as problem-solving and deriving general conclusions from specific experiments, than on knowledge (though even in traditionally knowledge-based subjects such as history, increased emphasis is now placed on mental skills such as analysis of different source materials). While it is true that the best students of science are always one step ahead of their teachers, no independent discovery of scientific principles is required up to research level. After all, a student would have to be as clever as Newton, and dedicate as much time to the task as he did, to come up with the laws of motion independently. At research level, the scientist's ability to apply knowledge and mental skills learnt from his predecessors and contemporaries must be supplemented by one additional ability that is not generally associated with science. It demands creativity. In experimental research, creativity is required to devise viable experiments to collect accurate data or to provide unambiguous evidence to support or challenge scientific theories. The need for creativity in theoretical research can be understood in terms of the impossibility of proving that a premise is true.

Mathematicians and scientists deal with this problem by coming up with a number of premises, provisionally assuming them to be true, and applying the rules of logic to arrive at conclusions on this assumption. These conclusions may be contradictory, in which case the premises are rejected as inconsistent. For example, consider the following premises:

1. All frogs have legs.
2. All frogs are animals.
3. All frogs live in water.
4. All animals that live in water lack legs.

From premises 2, 3 and 4, it can be deduced that all frogs lack legs. This conclusion directly contradicts premise 1, so one or more of the premises must be rejected.

Logic, then, can be used to test the consistency of a set of premises, but the question remains of how to establish the premises in the first place. Mathematicians are not concerned whether their ideas represent anything in the real world, and so are free to experiment with other-worldly premises leading to other-worldly conclusions. They are particularly concerned with axioms. An axiom is one of a number of premises that define a complete and consistent mathematical system with perfect economy. Any premise that could be deduced from the other premises would be excluded from the set of axioms for the system because it is unnecessary.

A particularly interesting example is the set of five axioms proposed by the Greek mathematician Euclid in the third century BC. Euclid was able to derive from these axioms the complete and consistent system of geometry still taught in schools today. Later mathematicians, however, were not satisfied merely with a set of axioms from which the geometry of the real world could be deduced. Euclid's fifth axiom states that two non-parallel straight lines meet if extended indefinitely, whereas two parallel straight lines do not (this is often stated as 'parallel lines meet at infinity'). This seems quite reasonable, but is suspicious in being more complicated than Euclid's other axioms. When nineteenth-century mathematicians investigated the consequences of rejecting the fifth axiom, they found that systems of geometry could be deduced from the other four which, while bearing no apparent relation to the real world, were nonetheless complete and consistent. It turned out that such non-Euclidean geometry does in fact have important applications in the real world. For example, the geometry of curved surfaces, such as the surface of the earth, is non-Euclidean, in that lines of longitude are parallel at the equator and straight from the point of view of someone on the surface of the earth, but nonetheless meet at the north and south poles. And, as proposed by Einstein in his general theory of relativity, the geometry of space itself is non-Euclidean on a large scale.

Scientists, in contrast to mathematicians, certainly are concerned with the real world. Scientists used to call their premises 'laws', but twentieth-century scientists, in recognition of the mutability of scientific ideas, have substituted the term 'postulates'. So Newton's laws of motion and the postulates of relativity are both examples of scientific premises. Such postulates are judged primarily by the accuracy with which conclusions derived from them match observations of the real world. (Other criteria by which they are judged include whether it is theoretically possible to make observations that prove the postulates wrong, and whether the postulates are simple.) The difficulty with finding such postulates is that, because premises cannot be deduced from conclusions, only conclusions from premises, scientific postulates cannot be deduced from scientific observations. So to call a world-view unscientific is hardly to discredit it, when all of science is based on postulates that cannot be proved true. (Scientific postulates do have the virtue that they can be proved false or contradictory, whereas people's world-views are often formulated so as to be unfalsifiable, and often contain conveniently overlooked contradictions.)

Philosophers of science have long pondered how scientists establish postulates if not by logical deduction, but they have failed to give a convincing account. All their attempts to describe the process of scientific discovery have either been peculiarly specific to a single instance of scientific discovery, or have borne little resemblance to what scientists do in practice. Their failure is not surprising, considering that the process of formulating a new scientific theory seems to have been different for every major scientific discovery of the past. (If it were the same every time, scientific discovery would be a simple matter of applying the same process in different circumstances.) Philosophers of science are finally conceding that analysis of the process of scientific discovery falls into the realm of psychology, not philosophy.

Psychologists have recognised scientific discovery as a creative process. Fundamental to any scientific discovery is the formulation of hypothetical postulates, which cannot be achieved either through logic or through any technique common to each of the scientific discoveries of the past. It demands creativity.

Neither logic nor science, then, provides the basis for human thought. They are merely disciplines that require the same processes of intelligence and creativity as other human activities. Similarly, neither logic nor science should provide the basis for computer thought, if the computer is to be intelligent in the same way as humans. The intelligent computer should acquire knowledge of logic and science and acquire the mental skills required in these disciplines in the same way as any other knowledge or mental skills.

Current Research

Since the invention of computers, considerable effort has been expended on reproducing human intelligence on a computer. In the early days of the electronic computer, some people believed that artificial intelligence would be quickly achieved, but the goal has proved elusive. Even attempts to reproduce those acts of intelligence that we take for granted, such as our ability to recognise a face, have met with only limited success. Any claims that computers could be as intelligent as humans must, then, consider how countless attempts to make intelligent computers have so far failed to achieve intelligence even approaching that of humans.

The description of commercial software that manipulates and applies information about the user as 'intelligent' is the understandable result of the over-enthusiasm of programmers and marketing agents. The description of many research projects as attempts at 'artificial intelligence' is less excusable. Projects such as the programming of a computer to perform logical analysis of sentences in the English language mentioned earlier are fascinating research projects that deserve funding. As attempts to achieve artificial intelligence, though, they are misguided, or, more likely, merely misnamed. Perhaps universities' preoccupation with departmentalisation would prevent such a project from gaining support if it were classified as a computing and linguistics project, rather than as an artificial intelligence project.

Research into the development of expert systems provides further examples of the misapplication of the term artificial intelligence. An expert system is a computer that emulates an expert of a given profession, such as a doctor, by encoding her knowledge on a computer. The computer assumes the role of the doctor by asking the patient a series of questions about his symptoms. Simple expert systems are programmed in a crude way, so that the programmer might ask the human doctor whose knowledge is to be encoded: 'What is the first question you would ask a patient if you could not see him?' The doctor might suggest: 'What is your main symptom?' and suggest possible answers: 'Headache, stomach ache, dizziness or faintness, burns or scalds, cuts, bruises' and so on. The programmer would duly provide the computer with instructions to ask the patient this question and allow him to choose one of the suggested answers. The programmer would then ask the doctor: 'If the patient answered that his main symptom was a headache, what would be your next question?' and then the doctor would suggest a follow-up question and another list of possible answers. Eventually, the programmer would be able to ask the doctor: 'Given the patient's answers, what would you consider to be the possible diagnoses?' The doctor would give a list of possible assessments of the patient's ailments, from the most likely to the least likely. The programmer would provide the computer with instructions for all these questions, answers and assessments, with the result that the computer could provide a patient with a reasonably accurate diagnosis. Most expert systems are more sophisticated than this, but all are based on these simple principles.

Such expert systems may appear to be intelligent, but they are not. Like so-called intelligent word-processing, educational and games software, an expert system learns information about the user, and manipulates the information it learns. But the ways in which the information is learnt and manipulated are not flexible, but fixed, predetermined by the programmer and the expert. The intelligence exhibited by an expert system is the combined intelligence of the programmer and the expert, not the computer. This is not to say that research into expert systems is not worthwhile, or that the resulting systems are not useful. But it is misleading to describe it as artificial intelligence research.

In the short term, research based on conventional programming, producing computers that act intelligently but are not intelligent, allows useful products to be developed quickly. In the long term, though, conventional programming is a particularly laborious way to make a computer act intelligently. The intelligent computer could be trained to perform a multitude of tasks, whereas a computer that has been programmed to act intelligently can only perform the task it has been programmed to perform, and only in the exact way it has been programmed to do it. For simple tasks, such as word processing, it is acceptable, indeed preferable, that the computer does only what it has been programmed to do, acting in the same highly mechanical way every time it is used. For tasks requiring more intelligence, such as the diagnosis of illnesses or the analysis of share prices, it is a considerable task to encode intelligent reactions to all possible eventualities. Further, each new application requires a new program to be written. In the long term, the most economical way to make a computer act intelligently will be to make it truly intelligent.

Research into neural networks represents a more enlightened approach to artificial intelligence. Neural networks are electronic circuits, or computer simulations of such circuits, that process information in similar ways to the neurons in the human brain. They have been used successfully to perform tasks such as reading handwriting, analysing statistics, controlling automatic machinery and recognising objects in photographs. However, it is again questionable whether neural networks could be truly intelligent.

A neural network is made up of a number of connected cells that act in much the same way as the entities of the associative model of memory. Each of the connections between the cells has a certain strength, according to which activation spreads from cell to cell. For example, consider a neural network designed to read handwriting (see Figure 16). A digital camera might be used to capture an image of the handwriting to be read. The pixels of a single letter of handwriting would then be fed into the input cells of the neural network. For the five-pixel-by-five-pixel image shown in the diagram, there might be twenty-five input cells, one for each pixel. For each black pixel, the corresponding input cell would be activated; for each white one, the corresponding input cell would not. The activation would then spread from cell to cell through the neural network according to the strengths of the connections between the cells. Finally, the activation would reach the output cells. For a neural network designed to distinguish the letters of the alphabet, there might be twenty-six output cells, one for each letter. If the connections between the cells of the neural network are of appropriate strengths, activation will spread

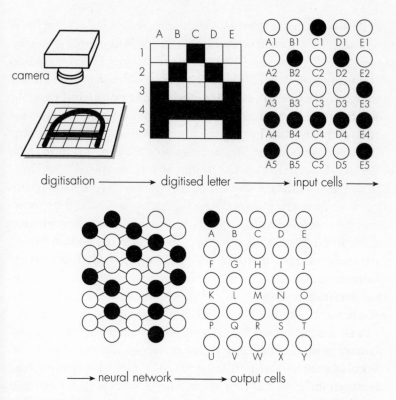

Figure 16. A neural network for recognising the letters of the alphabet.

to just one of the output cells, the one corresponding to the hand-written letter being processed. Once the neural network has recognised the letter, the pixels of the next letter of the hand-writing will be fed into the input cells. One letter at a time, the neural network will succeed in reading the handwriting.

The strengths of the connections in a neural network are not decided by a programmer. Instead, neural networks learn through training. Training a neural network to read handwriting is a slow process. The trainer might begin by capturing a digital image of the letter 'A' and feeding the pixels of the image into the input cells of the neural network, simultaneously activating output cell A. A feedback mechanism would then make small adjustments to the strengths of the connections between the cells of the neural network. Because the input cells A4, B4, C4, D4 and E4 are all activated by the crossbar of the 'A', the connections through the neural network between these cells and output cell A would be strengthened. So whenever a digital image of the letter 'A' is fed into the neural network in future, the activation of these input cells will be more likely to cause the required activation of output cell A, because the connections between them are stronger. Similarly, because the input cells C2, B3, C3 and D3 are all left unactivated by the enclosed area above the crossbar of the 'A', the connections through the neural network between these cells and output cell A would be weakened. So whenever a digital image of a letter that does not have such an enclosed area is fed into the neural network in future, the activation of the input cells will be less likely to cause the unwanted activation of output cell A.

The trainer might proceed by capturing a digital image of the letter 'B', feeding the pixels of the image into the input cells of the neural network, and simultaneously activating output cell B. The feedback mechanism would again make small adjustments to the particular strengths of the connections between the cells, this time reinforcing the connections between the input cells activated by the image of the letter 'B' and output cell B. Because these adjustments for the letter 'B' are small, the previous adjustments for the letter 'A' would not be obliterated.

The trainer would continue to present the neural network with inputs and outputs in this way. If the neural network is sufficiently complex, and if it has been trained with a sufficiently large number of letters, then it will be able to recognise the letters of the alphabet without any further prompting from the trainer. From then on, whenever a digitised image of a letter is fed into the input cells, activation will spread through the neural network in such a way that the output cell for that letter is activated.

In many ways, neural networks feel like the right approach to artificial intelligence. Their connected cells are analogous to the connected neurons of the human brain. They can cope with variations in input so that, for example, the shape of the letter 'A' does not have to be exactly the same every time for the neural network for reading handwriting to recognise it correctly. This behaviour is more reminiscent of the flexibility of the human brain than of the pedantry of a conventionally programmed computer. Further, neural networks generate confused output when presented with ambiguous input, which is again more reminiscent of human behaviour than computer behaviour. These analogies are enticing, but ultimately superficial.

A number of differences between neural networks and the human brain can be cited to counter the similarities. One difference is that humans can learn information immediately, whereas neural networks must be trained with a number of samples before the information is learnt. For example, if I invented a new letter of the alphabet, a human could learn to recognise it immediately, having seen it only once, but the neural network for reading handwriting would have to be trained with a large number of samples before recognising it reliably. Another difference is that the feedback mechanism, so fundamental to the operation of neural networks, simply does not exist in the human brain. For the feedback mechanism to operate, the trainer must specify the output required for each input. So when he feeds the digital image of the letter 'A' into the input cells of the neural network for reading handwriting, he must activate output cell A to specify that this constitutes the required output.

It is uncertain what constitutes a 'required' output for the human brain; indeed, it is questionable whether the concepts of input and output can be applied to the human brain at all. Moreover, a feedback mechanism requires that information from the output cells be communicated back through the neural network, so that the strengths of the connections can be adjusted according to the output specified by the trainer. Neurologists confirm that no such back communication of information takes place in the human brain.

These differences show that neural networks and the human

brain do not operate in the same way. But as I have stressed throughout this chapter, it is important not to assume that the human brain provides a perfect model of intelligence. The operational differences do not preclude the possibility that neural networks could be made as intelligent as humans.

There is, however, a more fundamental difference between neural networks and the human brain. A neural network has a trainer, and the trainer has an agenda. The role of the trainer of a neural network is not far different from the role of the programmer of a computer. In both cases, the human provides the machine with instructions. True, the instructions of the trainer are implicit and imprecise, whereas the instructions of the programmer are explicit and precise, but in both cases, the machine is instructed to do what the human specifies. In contrast, for the human brain there is no trainer, no agenda. Where neural networks are trained, humans learn.

There is an old story, perhaps apocryphal, of a neural network trained to spot tanks in military reconnaissance photographs. Photographs were digitised and the pixels of the images fed into the input cells of the neural network. There were just two output cells, one corresponding to the presence of a tank, and one corresponding to the absence of such a vehicle. The neural network was trained with hundreds of photographs. Whenever an image showing a tank was fed into the input cells, the trainer activated the output cell corresponding to the presence of a tank; whenever an image which did not contain a tank was fed into the input cells, the trainer activated the other output cell. However, when the trained neural network was tested with new photographs, it failed to work as intended, often indicating the presence of a tank in photographs that did not, in fact, show one, and vice versa. When the problem was analysed, it was discovered that, of the photographs with which the neural network had been trained, most of those that showed tanks were dark in tone, while a majority of those that did not show tanks were light. So all the neural network had been trained to do was recognise the difference between dark and light photographs! This story, whether or not it is true, illustrates the same point as the earlier accounts of the unintelligent behaviour of conventional computers provided with instructions that are incomplete or incorrect. When neural networks are provided with training that is inadequate or ill conceived, their behaviour is entirely unintelligent. Just as the intelligence of the conventional computer is that of the programmer, not the computer, so the intelligence of the neural network is that of the trainer, not the neural network.

Neural networks have attracted considerable attention. Their popular appeal lies in their simplicity and in the enticing analogy between the structure of a neural network and the structure of the human brain. Their research appeal arises perhaps from the complex mathematical analyses of neural networks which can provide an impressive basis for a research paper. There have been many successful practical applications of neural networks, though alternative solutions based on conventional programming have sometimes proved superior. As pure artificial intelligence research, neural networks are undoubtedly a step in the right direction, away from conventional programming and towards a more connectionist architecture. But it is a shame that the research establishment has not taken the further steps of abandoning the feedback mechanism and retiring the trainer.

An important problem with artificial intelligence research, including neural network research, is that its scope is often severely limited. Projects that concentrate on just one or two aspects of intelligence often produce unsatisfactory results, because *all* aspects of intelligence must be reproduced on a computer before true intelligence can arise. Consider, for example, a project to develop a computer that recognises faces, requiring the reproduction of visual perception on the computer. The researchers working on this project would probably approach it as the problem of matching a photograph of a face with a photograph in a database. When a human recognises a face, though, it is from a moving image rather than from a still image in a photograph. The additional information provided by the movement is invaluable in countering the confusing effects of seeing the face from different angles and distances, and under different lighting conditions. In limiting the scope of the project to the analysis of still images, the researchers have saved themselves the trouble of reproducing the more difficult analysis of moving images on the computer, but have made the task of face recognition considerably more difficult. Further, when a human recognises a face, it is in a particular context. Contextual cues, such as the manner of the person whose face is to be recognised, are vital to visual perception. Again, in limiting the scope of the project to encompass visual cues but no contextual cues, the researchers have saved themselves the trouble of developing a computer able to acquire the contextual knowledge required for the processing of contextual cues, but have made the task of face recognition more difficult. Of course, for commercial projects, the researchers would soon find themselves out of favour with their employers if they insisted that the development of a fully intelligent computer was a prerequisite to the completion of the face recognition project. For pure research projects, though, limiting the scope to one particular aspect of intelligence may yield quicker results, but will never allow the development of a computer as intelligent as humans.

Ultimately, there can be no half-measures in the reproduction of human intelligence on a computer. A computer that can only learn about its environment through language will never have the same understanding of a peach as humans who can see it and feel it and taste it. A computer that can see but cannot move will never have the same concept of distance as humans. A computer that has sensors for perceiving its environment but lacks the motors to manipulate it will never develop the same sense of self as humans. Any project to make a computer as intelligent as humans demands the reproduction of every aspect of human intelligence on the computer, freedom from stipulations that the computer be able to perform particular acts of intelligence (no right-minded parent would attempt to mould his child to become a mathematician or musician against the child's will), and the patience to wait for the completed computer to mature over a period of years (it takes decades for a human to mature). All too often, artificial intelligence researchers seem limited in their ambitions.

Unfortunately, the scope of artificial intelligence projects is limited not just by lack of ambition. It is also limited by technology.

Technology

Even if all the issues about the creation of the intelligent computer were fully resolved, there would remain formidable obstacles to making such a computer in practice. What a computer can do is limited not only by the set of instructions with which the programmer provides it, but also by its memory and its speed, and by its input and output devices. If the intelligent computer is to be made in practice, it needs to employ technology sufficiently advanced that none of these limitations has a significantly adverse effect on the intelligence of the computer.

The technology available for input and output devices is undoubtedly sufficiently advanced. For input devices, video cameras simulate the human eye, microphones simulate the ear, motion detectors simulate the sense of movement, and contact, pressure and temperature sensors simulate touch. If human senses of smell and taste are difficult to simulate, it is not a great loss to an intelligent computer less concerned with food than are humans. In compensation, the intelligent computer could be provided with input devices for detecting a wide variety of information unavailable to humans. For output devices, loudspeakers simulate human vocal cords, and robotic bodies simulate the human body. Robots currently available are not as agile as humans, and, because it is difficult to power them without cables, not as mobile. While this would no doubt be a source of frustration for the intelligent computer, it would not have a significantly adverse effect on its intelligence. For the designer of the intelligent computer, the limitations of the input and output devices are the least of his technology-related concerns.

The limitations of memory are more pressing. The standard unit of memory on a computer is the byte, which is the amount of memory required to store a single letter of a document. (So for this sentence, which contains 175 letters including spaces, figures, parentheses, commas and full stop, to be stored on a computer would require 175 bytes of memory.) The memory of a typical personal computer is measured in megabytes, one megabyte being one million bytes, or in gigabytes, one gigabyte being one billion bytes. For example, a personal computer might be equipped with a one-gigabyte hard-disk drive – enough memory to store the complete text of the Bible more than a thousand times over. The largest hard-disk drive widely available at the time of publication of this book has a capacity of some twenty gigabytes.

A computer modelling the human brain would need enough memory to record every neuron and every connection between neurons. Since the human brain contains some hundred billion neurons, each connected to some thousand other neurons, the computer would need a memory of the order of one million gigabytes. If this were to be provided in the form of twenty-gigabyte hard-disk drives, fifty thousand drives would be required. Even with connection techniques specifically designed to accommodate large numbers of devices, it would be difficult to connect so many. The expense of such an installation would be prohibitive.

The technophile's retort would be that, over the decades since the invention of the electronic computer, the capacity of memory devices such as RAM (Random Access Memory) chips and hard-disk drives has risen exponentially, and their cost has fallen exponentially. The designer of the intelligent computer need wait only another decade or two for a million-gigabyte memory to become feasible. Unfortunately, the fact that the capacity and cost of memory have varied exponentially in the past does not mean that they will continue to do so in the future.

Consider the future development of RAM chips. A RAM chip is an electronic circuit made up of millions of transistors etched on to a small wafer of silicon. The capacity and cost of a RAM chip depend primarily on the size of the transistors: the smaller the transistors, the more of them can be squeezed on to the same wafer of silicon. In the past, chip manufacturers have made impressive progress in reducing the size of the transistors. It is now possible to etch hundreds of millions of them on to a chip no bigger than a centimetre across, so that each transistor is less than a thousandth of a millimetre in size. Now, however, the manufacturers are coming up against limits that seem insurmountable.

The standard method of manufacture of silicon chips involves coating a silicon wafer with a light-sensitive film, exposing the film to light projected through a mask, then using chemicals to wash away only those parts of the film that were exposed to the light. The circuit is built up layer by layer, with a different mask for each layer, by selectively etching the silicon or introducing impurities into the silicon according to the template provided by the mask. One problem with this method of manufacture is that the mask must be the same size as the circuit itself. The smaller the transistors to be etched into the silicon, the more detailed the mask must be, and so the more difficult it is to manufacture. Another problem is that the more detailed the mask, the more the light is scattered as it passes through. The scattered light falls on to the wrong parts of the film and causes the wrong parts of the silicon to be etched.

Manufacturers have tackled the scattering problem by using ultra-violet radiation, which scatters less than visible light. Further improvements might be possible using X-rays, which are scattered still less. Alternatively, it might be possible to use electrons rather than electromagnetic radiation. One method would be to use the mask to scatter electrons rather than block them, transforming scattering from a problem into a solution. After passing through the mask, the electrons would be focused using magnetic fields so that the unscattered electrons would pass through an aperture and fall on the coated silicon wafer, while the scattered electrons would miss the aperture. Focusing the electrons in this way would have the further advantage that the mask could be made larger than the circuit to be etched, obviating the problem of manufacturing ever-smaller masks.

However, even the use of electrons to etch circuits on to silicon wafers might not allow transistors to be made significantly smaller than they are today. According to physicists, matter smaller than a certain limit, sometimes called the quantum limit, behaves according to the rules of quantum mechanics. These rules allow strange effects such as quantum mechanical tunnelling, whereby a particle such as an electron can penetrate an electrical barrier which, according to the more intuitive rules of classical mechanics, should confine it. Transistors are now so small that they are close to the quantum limit, and so could not be made much smaller without these strange effects interfering with their operation. The quantum limit might one day be surpassed by reinventing electronics to handle, or even to exploit, the peculiarities of quantum mechanics. It might then be possible to improve the capacity and cost of RAM chips significantly, and so to provide a computer with enough memory to model the human brain. It is impossible, though, to predict when the required leap in technology might be made.

If the limitations of memory are damaging to the prospects of making an intelligent computer, the limitations of speed are disastrous. The human brain is slow. It can take up to a tenth of a second for a signal to be transmitted between neurons, so the brain works on a timescale in the region of a tenth of a second. Computers, on the other hand, are fast, working on a timescale in the region of a billionth of a second. It seems odd, then, to claim that the limitations of a computer's speed are so damaging to the prospect of it being made as intelligent as a human.

However, the human brain has an all-important advantage over computers in that it works in parallel. While it takes as much as a tenth of a second for a signal to be transmitted between neurons in the human brain, billions of such signals are transmitted simultaneously. Computers, on the other hand, work serially. Although it takes as little as a billionth of a second for a computer to execute an instruction, only one instruction can be executed at a time. For most applications, the serial nature of computers is not apparent. For example, as a user types a document into a computer running word-processing software, the computer has to perform several tasks, such as checking the keyboard to determine what the user is typing, checking the mouse to determine if the user has clicked on a button on the screen, displaying any text the user has typed, reformatting and redisplaying any text displaced by the typing, and so on. The computer performs these few tasks in quick succession hundreds of times a second, so it is not apparent to the user that they are performed one after the other rather than simultaneously. However, for an application as complex as simulating the human brain, the serial nature of computers is paralysing. Instead of performing the few tasks required for word processing, the computer must simulate the behaviour of some hundred billion neurons in quick succession. If it takes, say, a millionth of a second to simulate the behaviour of a single neuron, it would take a hundred thousand seconds, a little over a day, to simulate the behaviour of all the neurons in the human brain. To give the appearance of working in parallel, the computer would have to perform this simulation hundreds of times a second, not once a day. It would have to be some ten million times faster than computers are today to achieve this. Not even the most avid technophile would claim this is likely to happen any time soon.

An alternative to increasing the rate at which a computer can execute instructions is to increase the number it can execute at the same time, in other words, to make the computer work in parallel, as the human brain does. Much research has been done into parallel computers, not just as a means to achieving the speed required for artificial intelligence, but as a means to increasing the speed of computers for all applications. The approach generally taken is to connect together a number of serial computers. For example, sixteen computers, each capable of executing a single instruction at a time, may be connected together, allowing up to sixteen instructions to be executed at once.

One problem with this approach is the difficulty of the efficient distribution among the sixteen computers of the instructions provided by the programmer. It is remarkably difficult to write a computer program in such a way that all sixteen computers are executing instructions at all times. All too often, the calculation to be performed by one of the computers depends on the result of a calculation performed by another, so that the first computer must lie idle until the second computer has finished. In practice, then, increasing the number of computers by a factor of sixteen increases the speed by a factor of somewhat less than sixteen. Another problem is that as the number of computers increases, the distance between them increases, so that communication between them becomes slower and more difficult to coordinate. Sixteen computers could feasibly be etched on to a single silicon chip, but sixteen thousand computers could not. Indeed, such a large number of computers probably could not be fitted into a space much smaller than about a metre across. It would take several billionths of a second to transmit a signal across so large a space, even if the signal were transmitted at the speed of light. In computer timescales, several billionths of a second is a long time.

So for parallel computers there is a trade-off between complexity and speed, but the relationship is lopsided. The greater the complexity of the parallel computer in terms of the number of serial computers connected together, the greater its speed. But as the parallel computer becomes more and more complex, ever smaller increases in speed are bought at the price of ever greater increases in complexity. To increase the speed of a parallel computer to that required to simulate the human brain, some ten million times faster than serial computers are today, would require a prohibitively expensive increase in complexity.

Another approach to parallel computing is more promising. Instead of programming a multi-purpose parallel computer to achieve artificial intelligence, the researcher could design a custom-made parallel computer. Purpose-built electronic circuits would implement the various functions of the intelligent computer. One circuit would be built to implement the initial stages of the processing of visual information. The electronic components would act in the same way as the spot-, line- and edge-detecting neurons in the human brain, the spot-detecting components combining signals from components connected to a video camera, and the line- and edge-detecting components combining signals from the spot-detecting components. Another circuit would be built to implement the associative model of memory, in which the electronic components represent entities and associations. The circuit would be designed so that associations would be strengthened through use, so that activation would spread from entity to entity through associations, so that the formation of those associations would be prompted by coincidence of activation of entities, and so on. Processes too complex to be implemented as purpose-built electronic circuits could be handled by a traditional multi-purpose computer interconnected with the custom-built circuits.

In a way, a custom-made parallel computer would not be a computer at all, in that it could not be programmed to perform any task. I will persist in calling it a computer, though, because the only real differences between designing a customised parallel computer and programming a multi-purpose parallel computer are those of convenience (a multi-purpose computer can be reprogrammed far more easily than a purpose-built machine can be redesigned and rebuilt) and speed. A custom-made parallel computer is clearly much faster than a serial computer because the electronic components of the former process signals simultaneously, whereas the latter executes a single instruction at a time. The customised parallel computer is also much faster because it operates without instructions. The function of the electronic components is determined by their structure rather than by instructions provided by the programmer, so the problem of the efficient distribution of instructions provided by the programmer is obviated. The custom-made parallel computer is also considerably less expensive than the multi-purpose one, because the former consists of only the electronic components required for the task, whereas the latter consists of the electronic components required to perform all possible tasks. Further, because it consists of fewer electronic components, the customised parallel computer can be squeezed into a smaller space, so the problem of the limited speed of transmission of signals across the computer is reduced.

The question of how such a computer might be manufactured is an interesting one. The most obvious way would be to etch the electronic components on to a wafer of silicon chip as described above. This is the standard method of manufacture not only for RAM chips, but for a diverse range of chips from audio amplifiers to microprocessors (a microprocessor is the main component of a serial computer). However, it might be possible to devise an entirely different method of manufacture for customised parallel computers. The standard method is based on the use of large, cumbersome tools to make small, precise circuits. It would perhaps be more appropriate to use small, precise tools the same size as the circuit to be manufactured, the size of molecules. The tools might take the form of complex molecules which encode the design of the circuit, and, when activated, effect its manufacture. One feature of custom-made parallel computers makes such molecular tools particularly suitable for their manufacture. Whereas microprocessor circuits are heterogeneous, every part of the circuit being different from every other part, the circuits of a custom-made parallel computer for artificial intelligence would be extremely repetitive. Each set of components in the visual processing circuit for spot detection would be the same, differing from others only in that it processes a different part of the computer's field of vision. Each set of components representing an entity in the memory circuit would be the same, differing from others only in its associations. Such repetitive structures might allow the design for a set of components to be encoded in a molecular tool, which would then be set to work to manufacture the set of components over and over again.

The idea that molecular tools might be used to manufacture electronic circuits may seem far-fetched, but such tools are common in nature. Indeed, all living things are manufactured using such tools. Humans are no exception. DNA molecules in our cells encode the blueprint for the human body, and RNA molecules effect the manufacture of the human body according to this blueprint. This method of manufacture is viable because, on a small scale, the structure of the human body is extremely repetitive, every cell having the same basic structure as every other. Of course, there is some specialisation among cells (a neuron is different from a red blood cell) and among different parts of the body (an eye is different from a leg), and it is to encode such microscopic and macroscopic variations that human DNA must be so complex. But the basic principle according to which the human body is manufactured is the use of molecular tools to give a repetitive structure.

The same principle might one day be applied to the manufacture of customised parallel computers for artificial intelligence. It seems likely that carbon, the element on which all living organisms currently known to scientists are based, would prove to be a more appropriate basis for this manufacturing process than silicon. If so, the intelligent computers of the future might prove to be remarkably similar to the human brain in composition as well as in structure.

The technology required to make custom-built parallel computers using molecular tools is not currently available, nor is it likely to become so in the immediate future, but this does not doom to failure all attempts to reproduce intelligence on a computer in the meantime. In the absence of such technology, two options remain. One is to use existing chip-manufacturing technology, but the drawback of this option is the expense. For commercial ventures, such as the manufacture of a microprocessor, current chip-manufacturing technology is cost-effective. The initial research and development costs involved in the manufacture of a microprocessor chip are high, but once the design is finalised, the cost of manufacturing each chip is relatively small. Because the demand for microprocessors is considerable, the initial research and development costs are easily recovered. However, for research projects, such as the manufacture of a customised parallel computer for artificial intelligence, the high initial costs may never be recovered. The other option is to revert to using slow, but comparatively cheap, serial computers.

The expense of customised parallel computers and the slowness of serial computers are not as damaging to artificial intelligence research as might be imagined. Just as for multipurpose parallel computers there is a trade-off between complexity and speed, for the intelligent custom-made parallel computer there would be a trade-off between complexity and intelligence. The greater the complexity of the specially designed computer, the greater the scope for intelligence. The more spot-, line- and edge-detecting neurons simulated by the computer's visual perception circuit, the more accurately the computer would be able to resolve objects. The more entities and associations represented by the computer's memory circuit, the more knowledge the computer could acquire. So limited resources and limited technology, in restricting the complexity of the customised parallel computer, would constrain the intelligence of the computer, but would not preclude intelligence altogether. As technology improved, increasingly complex customised parallel computers, and therefore increasingly intelligent ones, could be manufactured. It is worth emphasising, however, that while complexity is necessary for intelligence to arise, it is not in itself sufficient. A badly designed computer will never be intelligent, no matter how complex it is.

For the intelligent serial computer, there would be an additional trade-off between speed and intelligence. If it were to take more than a day for a serial computer to simulate all the neurons in the human brain, such a computer could be as intelligent as a human, but would be so slow that it would seem inappropriate to describe it as intelligent. If it takes seconds for a human to understand a sentence spoken in conversation, it would take the computer months. Worse, if it takes decades for a human to achieve intellectual maturity, it would take the computer hundreds of millions of years. The intelligent serial computer could be made to operate more quickly by having it simulate fewer neurons, but the smaller the number of spot-, line- and edge-detecting neurons simulated, and the smaller the number of entities and associations represented, the less intelligent the computer, as with customised parallel computers. So again, limited resources and technology, by restricting the speed and therefore (if it is to operate at something like human speeds) the complexity of the serial computer, would thus limit its intelligence, but would not preclude intelligence. Again, as technology improved, increasingly fast serial computers, and therefore increasingly intelligent serial computers, could be manufactured.

So current technology, while not sufficiently advanced to allow a computer as intelligent as a human to be made at reasonable cost, nonetheless allows worthwhile research into artificial intelligence to be conducted. The computers resulting from such research will initially appear quite dull-witted, but the research will be invaluable for proving the principles of artificial intelligence. As technology advances, the computers will become increasingly intelligent. Eventually, perhaps with the advent of molecular tools for the manufacture of custom-built parallel computers, technology will advance sufficiently so that the computers *will* become as intelligent as humans. Indeed, the time will come when technology is so advanced that intelligent computers will take fractions of a second to have thoughts that would take a human many seconds. Then it will be humans who seem dull-witted.

The Intelligent Computer

If various aspects of human intelligence were reproduced on a computer as discussed, from sensation and perception to the learning of knowledge and concepts and skills, would it truly be as intelligent as a human? Or, like the unintelligent computer running word-processing software, would it merely give the appearance of intelligence?

Alan Turing, one of the pioneers of computer theory, suggested a test to determine whether a computer is intelligent, now known as the Turing Test. He imagined a human interrogator communicating with either a computer or a human in another room. The communication would take the form of messages carried between the rooms by a neutral messenger. These messages would be printed, so that only the content of the messages could give the interrogator clues as to whether he is communicating with the computer or the human. According to the Turing Test, if the interrogator is unable to distinguish the computer from the human, then the computer is as intelligent as the human.

Unfortunately, this test has many shortcomings. If the interrogator is untrained, it is easy to dupe him into believing that a completely unintelligent computer is as intelligent as a human. In one experiment, set up in much the same way as the Turing Test, unsuspecting subjects conversed with a computer about personal matters. The computer perpetuated the conversation by interjecting such prompts as 'How do you feel about that?' and 'Tell me more.' Sometimes, it would use words spoken by the subjects in its replies, so that if a subject said: 'I've had problems getting on with my father lately,' the computer might reply: 'Tell me about your father.' Many of the subjects were convinced that they had been talking to a human, and were impressed by the empathy they imagined had arisen between them. Even after it had been explained that they had been conversing with a computer, some of the subjects refused to believe it.

Another shortcoming of the Turing Test is that it requires the computer to be exactly like a human. People from the same culture grow up in a similar environment and acquire similar background knowledge from parents, from school and from the media. As a result, there is a remarkable uniformity in the basic knowledge expected of people (though, of course, every person has special knowledge and particular quirks that make him unique). An intelligent computer that has not grown up in the same environment as a human (it is difficult to imagine an intelligent computer as a pupil in a primary school, sitting in lessons and running in the playground with the children) would not have the same basic knowledge. This does not mean that the computer is less intelligent than humans, only that its experience is different from that of a typical human. Nonetheless, with an alert interrogator, the computer would fail the Turing Test.

These shortcomings are indicative of the more fundamental problem with the Turing Test. It treats the computer as a 'black box', a term applied by scientists and engineers to an object whose inner workings are unknown, so that it can be understood only in terms of its interactions with its environment. According to the Turing Test, a computer's intelligence is to be determined only in terms of its interactions with the interrogator, and not by considering its inner workings. This view has parallels in a movement in psychology in the first half of the twentieth century called Behaviourism. Behaviourists rejected the attempts of psychoanalysts such as Freud to understand the human brain through introspection and through analysis of subjects' reports of their own experiences, and were concerned only with scientific observations and measurements.

The Behaviourists introduced to the field of psychology a scientific rigour that had previously been lacking, but their refusal to consider the possibility of inner mental activity led to a tendency to treat the human brain as a black box. The strictest Behaviourists considered that every response exhibited by humans must be the result of an identifiable stimulus, rather than the result of inner mental activity, and that theories of the inner workings of the human brain should therefore be rejected as unscientific. Such extreme views were soon rejected by psychologists, because theories of the inner workings of the human brain turned out to be extremely useful in explaining human behaviour, and because neurological research began to allow such theories to be subjected to more rigorous scientific analysis. The human brain is not a black box, and its inner workings can be known.

Similarly, the inner workings of a computer can be known. It is perverse, then, to attempt to determine whether a computer is intelligent as if it were a black box. By considering its inner workings, we are better informed to answer such questions as whether the computer is truly as intelligent as a human or is merely composing responses that mimic human responses, whether the differences between the computer and a human are rooted in experience or are innate, whether the intelligence exhibited by the computer is that of the programmer or the computer. Just as Behaviourism, in denying the inner workings of the human brain, failed to provide a full explanation of human behaviour, so the Turing Test, in ignoring the inner workings of the computer, is not an adequate test for computer intelligence.

The intelligent computer must be able to learn information. The learnt information must be flexible, in that it can be manipulated and applied and used as a basis for further learning by the intelligent computer. The manipulation, application and use of the information as a basis for further learning must themselves be flexible, in that the intelligent computer can learn new ways to manipulate and apply information, and learn new ways to learn.

These stipulations would be met by a computer designed as I have suggested. A computer could be made to detect information about its environment through sensors, derive meaning from the information by processing it, and reduce the information by selection on the basis of importance. This information could be encoded in a flexible way as associations between entities, and recalled through the activation of the entities. Finally, the computer would acquire skills for the manipulation and application of this information. It would satisfy all the above criteria for intelligence.

It is important to clarify the origin of the intelligence exhibited by a computer. In my discussion of so-called intelligent software, I described how the programmer decides, using his own intelligence, what would be the most intelligent way for the computer to react to different circumstances, and provides the computer with precise instructions to react in this way, and how the computer mindlessly follows these instructions. I claimed that in this case, the intelligence is that of the programmer, not the computer. I went on to describe how a computer could be programmed, either explicitly in the instructions for a serial computer or implicitly in the design of a customised parallel computer, to reproduce the various aspects of human intelligence. I am now claiming that, in this case, the intelligence is that of the computer, not the programmer. These two apparently contrary claims must be reconciled.

Whether the intelligence of my computer model is that of the computer or the programmer is closely analogous to the question of whether the intelligence of a human is that of the human or of evolution.

The structure of the human brain allows an intelligence to arise that is probably unsurpassed in the natural history of the earth. It seems reasonable, then, to claim that the human brain is extraordinarily well designed. As is often the case with statements framed in our concept-bound language, though, this claim is misleading, because it implies the existence of a designer. On one level, the structure of the human brain is determined by our genes, but it is misleading to ascribe the role of designer to simple strands of DNA, which, through RNA, effect the manufacture of protein molecules in complete ignorance of their contribution to the structure of the brain. On another level, the structure of the human brain is determined by evolution. Billions of years of random mutation and natural variation, of organisms well adapted to their environment surviving and organisms less well adapted to their environment perishing, have accidentally culminated in the myriad of creatures alive today, among them humans with our apparently well-designed brains. But again, it is misleading to ascribe the role of designer to evolution, since it proceeds by accident, not by design.

If it is misleading to think of evolution as a designer, it is nonetheless true that evolution is responsible for the structure of the human brain, the structure that allows our intelligence to arise as we accumulate experience of our environment. However, the structure of the human brain does more than merely enable our intelligence; it also shapes our intelligence. Inherent in this structure are various hints and prejudices that direct and delimit the development of our intelligence. An example of such a hint is the crossing of the optic nerves in the human brain. The optic nerves from the left and right eyes cross so that the neurons from the left side of each eye end up in one half of the brain and the neurons from the right side of each eye end up in the other half. This allows the positions of features in the fields of vision of the left and right eyes to be directly compared in the same part of the brain, so providing a valuable cue for distance determination as discussed earlier. We learn for ourselves the correlation between this and the other cues for distance determination, so our concept of distance is a product of our own intelligence, but by facilitating the detection of this cue, the structure of the brain provides a useful hint.

An example of a prejudice inherent in the structure of the human brain is the interpretation of visual information as spots, lines and edges. Such an interpretation is appropriate to our environment, which consists of discrete objects whose sharp boundaries appear as lines and edges. But it would be entirely inappropriate in an alien environment consisting of nebulous clouds of gas appearing as subtle variations of colour and intensity without distinct boundaries. In this respect, then, the structure of the brain is well adapted to our environment, but consequently prejudiced in its perception of that environment.

Evolution, then, is responsible for the structure that gives rise to and shapes our intelligence. Beyond the provision of this basic structure, our intelligence is our own. Evolution provides us with a small amount of innate knowledge, or instincts – such as that a full stomach is conducive to survival – but otherwise we acquire our own knowledge. Evolution provides us with a small number of innate skills, or reflexes – such as withdrawal from pain, or a baby's crying in response to hunger – but otherwise we develop our own skills. Finally, evolution provides us with a small number of hints as to the nature of our environment – such as the hint that comparing the positions of features in the fields of vision of the left and right eyes provides a valuable cue for distance determination – but otherwise we form our own concepts. Evolution performed the research and development for the human brain, and our genes mastermind its manufacture, but all this served only to provide an organ capable of intelligence. The intelligence that subsequently arises in that organ is that of the human, not of evolution.

The analogy with the making of an intelligent computer is clear. The programmer of an intelligent computer designed as I have suggested is responsible for the structure that gives rise to and shapes the computer's intelligence. He provides the computer with a small amount of innate knowledge, a small number of innate skills, and a small number of hints as to the nature of the computer's environment. He creates a computer that is capable of intelligence, but the intelligence that subsequently arises in that machine is that of the computer, not the programmer.

The contrast with a computer programmed in the conventional way is striking. Whereas the conventional programmer decides, using his own intelligence, what would be the most intelligent way for the computer to react to different circumstances, the programmer of an intelligent computer does not even consider what circumstances the computer might encounter. Whereas the conventional programmer provides the computer with precise instructions that tell it what to do in all possible circumstances, the programmer of the intelligent computer provides it with a structure that allows intelligence to arise, so that the computer can decide for itself what to do in different circumstances. Whereas the set of instructions provided by the conventional programmer is an encoding of the programmer's knowledge, concepts and skills, the intelligent computer acquires its own knowledge, concepts and skills.

A computer designed as I have suggested, then, would possess an intelligence equal to that of humans, an intelligence that would be that of the computer, not the programmer. But what about other aspects of human intelligence that cannot be reproduced on a computer? Intelligence has so far been discussed only in its narrowest sense. The following chapters examine creativity, emotions and consciousness and propose that intelligence, creativity, emotions and consciousness are interdependent, and that none of these human traits could be reproduced on a computer independently of the others. A computer could not be truly intelligent in the same way as humans without also being creative, emotional and conscious in the same way as humans.

2

THE CREATIVE COMPUTER

Creativity

We describe a multitude of human activities as creative. Painting or sculpting, composing a symphony or writing a novel, designing a building or a bridge, inventing an electronic device or formulating a scientific theory – each is considered a creative activity, requiring different knowledge and different skills, but, in addition, each requires a special ability, creativity, that sets it apart from other activities. How are humans creative, and how could human creativity be reproduced on a computer?

A definition of creativity is required. Clearly, for an activity to be creative, something must be created. The creation need not be as tangible as a sculpture or a bridge. When a composer writes a symphony, he generally puts it on paper as a musical score. Only later, at the symphony's premiere, is it transformed into the audible form the composer originally intended. But the concert is not the creation, since the composition of the symphony would have been an act of creativity even if the symphony had never been performed. Nor is the score the creation, since the composition would have been an act of creativity even if the composer had composed the symphony entirely in his head without ever putting it down on paper. The creation is the symphony itself, intangible but nonetheless real.

Similarly, when a scientist formulates a theory, he describes it in a research paper, but the theory exists independently of the research paper, in the mind of the scientist who formulates it and in the minds of others to whom he communicates it. The creation is the theory itself, again, intangible, but again, real.

For an activity to be creative, not only must something be created, the creation must also go beyond the previous experience of the creator. A sculptor uses a hammer and a chisel to work on a block of stone in the same way whether she is making an original sculpture or an exact copy of another sculpture. Nonetheless, if she is making an original sculpture, she is being creative, but if she is making a copy, she is not. The original sculpture goes beyond her previous experience, the copied sculpture lies within her previous experience. Creations are rarely identifiable as either entirely beyond the previous experience of the creator or entirely within the previous experience of the creator. An architect may draw on standard techniques and traditional styles in her design for a building, so that the design is largely based on existing architecture, but if she combines these techniques and styles in an original way, or makes original use of a standard technique, or introduces an original variation on a traditional style, then she is still being creative.

For a creation to go beyond the previous experience of the creator, it need not go beyond the previous experience of anyone else. Suppose that two scientists independently formulate the same theory. With an increasing number of people engaged in scientific research in modern times, it is common for scientists from all over the world to investigate the same phenomena, so it is quite possible that two scientists may formulate the same theory at the same time, without any communication between them until each has completed his own version of the theory. Under such circumstances, it is absurd to suggest that the scientist who finished formulating his theory first has been any more creative than the other. Clearly, both have been creative. Even if, many years later, a third scientist formulates the same theory without any knowledge, direct or indirect, of the work of the first two, he too has been creative, because the theory goes beyond his previous experience. For an activity to be creative, then, the creation need not precede all similar creations in time.

Describing creativity in terms of the previous experience of the creator means that there can be no objective measure of creativity. Judges for awards such as the Nobel prizes for science have difficulty enough determining whether a creation precedes all similar creations in time. It would be impossible to determine whether a creation goes beyond the previous experience of the creator. Even the creator himself cannot know his mind so well that he could determine the extent to which an idea is his own, and the extent to which it is based on ideas from long-forgotten sources. Describing creativity in terms of the previous experience of the creator also means that there can be no objective comparison of acts of creativity. There is no way to compare a picture painted by a five-year-old child with one painted by a professional artist and decide objectively which is furthest beyond the previous experience of its creator. It would be impossible to analyse the previous experience of the child and the adult, encompassing everything they have ever seen, heard, thought and felt. It would also be impossible to define a measure of the extent to which the paintings of the child and the adult are their own, and the extent to which they are derived from these previous experiences.

I have described creativity as requiring that something be created and that the creation go beyond the previous experience of the creator. Something is missing from this description. I am not a painter, but I know that if I acquired brushes, paint and canvas, I could create something beyond my previous experience. For example, I could create my own style of painting in false colours – red for green, blue for yellow, light for dark – so that my paintings resembled negatives for colour photographs. No doubt someone with knowledge of the history of art would point out that such a style would be a shallow echo of some long-forgotten movement, but, as far as introspection allows me to ascertain, the style goes beyond my previous experience. I would be loath to describe painting in such a style as creative, though, because the paintings would simply be no good.

For an activity to be creative, the creation must be considered good, either by the creator or by the beholder. This does not mean technically competent. A good artist must be competent in the use of her preferred media and the execution of her preferred styles, but an artist may acquire technical mastery of her art and never produce a good work. A technically competent work of art may inspire admiration for the artist, but a good work of art stimulates different feelings in the beholder. A good painting, for example, might generate interest in the scene, in the place and the time portrayed. It might arouse curiosity about the circumstances in which it was painted, the life of the artist and the lives of the people in the painting. It might stir up memories of the past, and so evoke feelings associated with those memories. It might inspire compassion, wonder, joy, anger, fear, despair, possibly elicited through empathy for figures in the painting whose faces betray such emotions. It might give a sense of heightened perception, of the shadows under a tree, or of the folds in a dress, or of the form of a body, if these are painted with sensitivity. It might give a sense of altered perception by depicting light or motion or form in an unconventional way, and so presenting an alternative view of reality. Or it might simply generate the deep and satisfying sensation one experiences when looking at a beautiful creation. A good work of art inspires such feelings; a bad work of art inspires none.

Different people have different ideas of what is good. An artist might consider one of her paintings to be her greatest to date. An art critic might consider the same painting to be worse than the artist's previous work. Exhibited in an art gallery, the same painting will draw different reactions from different visitors: some will find it beautiful, others ugly, or fascinating, or dull.

Each person will have arrived at his assessment of the value of the painting in a different way. One visitor may have immediately found the painting ugly, and walked straight on to the next. Another may have instantly assessed it as beautiful, and spent some time looking at it, exploring its colours and shades. A third art-lover may have been impressed by the mood of the painting and studied the information supplied by the gallery about the artist, relating this information to the painting itself. The critic, too, will have formed an immediate impression of the painting, and connected it with what he knows about the artist, her previous work and her current circumstances. In addition, he will have analysed the style and technique, comparing them to those of other artists, assessed the skill of the artist, and speculated as to what the artist might have wanted to convey.

With his experience of analysing works of art in this way, the art critic is perhaps better able to explore the various facets of the work and to develop his initial impression of the painting. He may also be better able to verbalise and rationalise his view of the painting. It could be argued that the critic's abilities make his assessment of the value of the painting more valid than the assessments of the art gallery visitors. It could equally be argued that the visitor who explored the colours and shades of the painting, innocent of intellectual considerations of style and technique, is closer to the painting than the critic, so that her assessment of the value of the painting is more valid than the critic's. Such arguments are futile. No one's assessment of the value of a work of art is any more valid than anyone else's. What is good is entirely subjective.

Again, describing creativity in terms of whether the creation is good means that there can be no objective measure of creativity. There is no way to compare a picture painted by a five-year-old child with one painted by a professional artist, and decide objectively which is better. Suppose a visitor to an art gallery were asked which was the best work: a painting in the gallery generally recognised as a masterpiece, her own amateur pen-and-ink sketch of a church, or her five-year-old son's painting of a person (see Figure 17 on page 112). She might give the conventional answer that the masterpiece is better than her sketch, and her sketch better than her son's painting. But her true feelings about the three works might suggest that the order

Figure 17. A five-year-old's painting of a person, *Vase with violet irises against a yellow background* by Van Gogh, and an amateur pen and ink sketch of a church. Which is the best work?

be reversed. The masterpiece may have impressed her with its virtuosity, but her own unique evocation of the beauty of the church in her sketch may have given her greater reward, and the sight of the uninhibited brush strokes of her son's painting may have given her still greater pleasure. Of course, the value she places on her own sketch and her son's painting is inflated by her special interest in herself and her son. But it is impossible to decide objectively which of the three works is better. All that can be said about the three works is that one person has one view of them, another person has another view of them, and so on. For the art gallery visitor, and for her alone, her son's painting is of greater value than her sketch, and her sketch is of greater value than the masterpiece. How good something is, and so how creative something is, is entirely subjective.

For an act to be creative, then, something must be created, the creation must go beyond the previous experience of the creator, and the creation must be good.

The Uncreative Computer

A number of research projects – some initiated by artists, some by psychologists, and some by programmers – have aimed at reproducing creativity on a computer. These projects have produced computers that compose music, draw pictures (see Figure 18 on page 114), write stories, even formulate scientific theories. These computers have been programmed to act creatively, but they are not creative.

Consider how a computer might be programmed to compose music. An obvious first attempt would be to program it to generate random notes, each with a random pitch and a random duration. The resulting composition would be original in the sense that the precise combination of random notes generated would almost certainly never have been heard before. Any discussion of whether the composition would go beyond the previous experience of the creator merely raises the question of whether the computer can be said to have experience of its own. These considerations aside, the computer's composition

Figure 18. A drawing by a computer programmed to act creatively.

would sound dreadful. To call it music, let alone a work of creativity, would be to stretch the imagination.

The next step might be to program the computer to conform to the basic rules of music. The pitches of the notes would be confined to the pitches of the strings of a well-tuned piano, and further confined so that the music is in a particular key. The durations of the notes would be chosen so that the music divides into bars according to a particular time signature. The computer would be programmed with the rules for forming chords, the rules of harmony. These are simple, mechanical rules, the kind that set a schoolgirl daydreaming in her music lesson, but they present the programmer with no difficulties. The result of a random generation of music according to these rules would not be as grating as the previous attempt, but it would still not sound good. It would remain a random collection of notes, with no connections between the notes, no phrases, no themes.

The computer could be programmed with more and more rules: perfect cadences, imperfect cadences, interrupted cadences, plagal cadences; six-four chords, mediant chords, sub-dominant chords; melody, modulation. It could be programmed to create random phrases according to these rules, and work them into random themes with repetitions and variations. The result would become more and more palatable, but it would still be dull.

Programming further rules would only compound the problem. A computer programmed with rules formulated by the composers of the past would merely replicate their work in a mechanical way, rather than create something original. The most dramatic developments in music in the past have come from breaking the old rules and creating new frameworks for composition. Even within a single composition, the best moments often occur when the rules are stretched to breaking point.

The next step, then, might be to program the computer to stretch and break the rules, to develop its own framework for composition. The trouble is that the computer does not know what sounds good. It would be unable to differentiate between a change of rules that adds an exciting new element to the music, and one that simply makes it sound dreadful. If the computer were left running for long enough, it would eventually come up with a composition that would sound good to humans, but only in the same way that a large number of monkeys typing at random would eventually come up with the complete works of Shakespeare. The computer, like the monkeys, cannot distinguish the good from the mediocre, the mediocre from the downright awful.

The attempt to program the computer to compose music is founded on the implicit assumption that there exists an objective ideal of good music, and that it is possible to define a set of rules for its creation. This assumption is false. As suggested above, 'good' is entirely subjective. Some people like renaissance music, some baroque, some classical, some romantic, not to mention pop, rock, jazz, blues. What is surprising is not that humans have such a variety of different tastes, but that there is such uniformity. Among those who like Baroque music, a large number like Bach. Among those who enjoy pop music from the sixties, a large number are fond of The Beatles. It is tempting to conclude that Bach and The Beatles were objectively good. Certainly they were masters of their arts, but a composer may master his art and still lack popular appeal. Bach and The Beatles have such wide appeal not because their music is objectively good but because it seems to strike some common chord deep within human nature. Humans, particularly humans brought up in the same culture, are remarkably homogeneous in their notions of what is good, but a large number of consistent subjective notions do not add up to an objective truth.

For the computer to be creative, its creations must be good. The attempt to program the computer to compose music was misguided, because it failed to take this into consideration. Perhaps, then, the computer must be programmed with some notion of what music is good before it can be programmed to compose music that conforms to that notion. This raises the question of what notion of good the computer should be programmed with. It cannot be programmed with an objective notion of what music is good, because no such objective notion exists. This leaves a number of options.

One solution is to program the computer with a random set of criteria describing what music is good. In this case, though, the computer's notion of what is good would differ from that of humans. The computer could create a piece of music that it considers the pinnacle of its achievement, but a human listener would consider it to be dreadful. According to the description of creativity developed earlier in the chapter, this does not make the computer's composition any less creative, since the computer's assessment of the value of the music is no less valid than the human's. Indeed, many a modern composer has written a piece of music which he considers to be of value, but which the majority of the concert-going public considers to be dreadful (though they may later come to appreciate it). Nonetheless, it can hardly be classed as a success to have programmed a computer to compose music that only it considers to be good.

Another option is to attempt to identify the characteristics of the music that humans generally consider to be good, such as that of Bach and of The Beatles, and to condense these characteristics into a set of well-defined criteria with which to program the computer. It would then have a notion of what music is good that coincides with that of humans in general. However, quite aside from the practical difficulties of such an undertaking, I do not think it possible to condense what humans consider to be good into a set of well-defined criteria. What a human individual considers to be good, let alone what humans in general consider to be good, defies definition.

The real problem with programming the computer with a notion of good music lies in the programming. We have been talking about programming throughout – first programming the computer with rules for composition, then to stretch and break the rules, now to conform to a particular notion of what is good. This recalls the previous assertion that if a computer is programmed to act intelligently, the intelligence is that of the programmer, not the computer. In the same way, if a computer is programmed with rules for the composition of music, these rules represent the experience of the composers of the past, as collected and formalised by the programmer, not the experience of the computer. Any suggestion that music composed by the computer goes beyond its previous experience is meaningless when considered in this light, because the computer has no experience of its own. Further, if a computer is programmed with the programmer's notion of what is good, the notion is that of the programmer, not the computer. Again, the suggestion that the computer considers the music it composes to be good is meaningless, because the computer has no notion of good of its own. Creativity cannot be programmed.

Consider the research projects that have produced computers programmed to compose music, draw pictures, write stories, and formulate scientific theories. The claims that these computers are creative are false. Each of the computers has been programmed to follow rules, and to stretch and break these rules in ways defined by the programmer, based on the programmer's experience and notion of what is good. The computer's creations may go beyond the previous experience of the programmer, but the computer's only contribution to the creative process is randomness, which could just as well be made by rolling dice (if it were not for the fact that a computer can generate millions of random numbers in the same time it takes to roll a die just once). If the programmer considers the computer's creations to be no good, he will simply adjust the rules for creation, until the computer creates something that he does consider to be good. Even then, the computer may create many compositions or pictures or stories or scientific theories, from which the programmer selects the few that he likes. Always, it is the programmer's experience and the programmer's notion of what is good that defines the creation. The computer can be invaluable for exploring the possibilities for creation, but it is no more than a tool. The creativity belongs to the programmer, not the computer.

Deciding how a computer could be made truly creative requires a deeper understanding of the processes of creativity, and the drawing of a loose distinction between analytic and aesthetic creativity.

Analytic and Aesthetic Creativity

Creations in the fields of science and technology are generally evaluated analytically. Consider a physicist investigating collisions of protons in a particle accelerator. (All matter consists of subatomic particles such as protons, neutrons and electrons, and physicists have developed a penchant for accelerating these particles to high speeds, then causing them to collide.) The physicist's aim is to formulate a theory that accurately models the collisions of the protons, and her creation can be evaluated analytically. If analysis of her theory reveals that it models the collisions accurately, then the theory is good; if analysis reveals that the theory fails to model the collisions accurately, then it is no good. Consider an engineer drafting a design for a bridge between a large island and the mainland. The engineer's aim is to design a bridge that carries a certain load, perhaps one hundred cars at a time, over a certain distance, perhaps several hundred metres. In addition, the bridge must be able to withstand adverse weather conditions, such as high winds. Again, the engineer's creation can be evaluated analytically. If analysis of his design reveals that the bridge would carry the required load over the required distance with the required reliability, then the design is good. If, instead, the analysis reveals that the bridge would collapse in high winds, then the design is no good.

Creativity for which the creation can be evaluated analytically is what I will call 'analytic creativity'. Both the physicist and the engineer exercise analytic creativity.

It is worth emphasising that if the physicist and the engineer fail to meet the demands of their chosen fields, it does not necessarily mean that they have failed to be creative. If the physicist's theory is beautiful but does not accurately model the collisions of the protons, it will be rejected by other scientists, but a description of it might make fantastic reading in a science fiction novel. Similarly, if the engineer's bridge is beautiful but would collapse in high winds, it will be rejected by whoever commissioned it, but a painting of it might make a fantastic backdrop for a film.

It is also worth emphasising that the fact that the evaluation of the physicist's theory or the engineer's design is analytic does not mean that it is objective. The criteria according to which the engineer's design is evaluated are dictated by human requirements. The bridge must carry one hundred cars at a time because this is the load that humans want to move. If humans were content to walk across the bridge rather than take their cars, the criteria according to which the design is evaluated would be different. Similarly, the bridge must carry the cars between the large island and the mainland, because these are the points between which humans want to travel. If humans wanted to visit a small island just one hundred metres from the mainland rather than the large island several hundred metres from the mainland, then the criteria according to which the design is evaluated would again be different. The evaluation of the engineer's design according to specified criteria is objective, but the criteria themselves are subject to human whim.

Less obvious is the fact that the criteria by which the physicist's theory is evaluated are also dictated by human requirements. Science is a human construction, one way of understanding the universe, albeit a way that affords us the power to manipulate the universe through technology. It is to satisfy our desire to understand the universe according to the scientific world-view, and to increase our power to manipulate the universe through technology, that we insist that a scientific theory meet certain conditions. We insist that the theory be predictive, simple and falsifiable. We further insist that the phenomena it describes be measurable and either reproducible or frequently occurring. These criteria are dictated by human requirements. If a theory did not meet the criteria, it would not increase our understanding of the universe, nor would it increase our power to manipulate the universe, so it would be less valuable to humans. But it would not be less valuable in any objective sense.

Analytic creativity is not confined to science and technology. Creations in art, too, can be evaluated analytically. An art critic may evaluate a painting by analysing the artist's manipulation of the brushes and paint, or the interplay between the different elements of the painting, or the influences of various art movements on the development of the artist's style. Again, it should be emphasised that just because the art critic's evaluation of the painting is analytic, it does not mean that it is objective. If the artist has been deliberately naive in her manipulation of the brushes and paint, or if she has deliberately played down the interplay between disparate elements of the painting, or if she has departed from the conventions of current art movements, it does not mean that the painting is no good.

There is an important sense in which the analytic evaluation of the artist's painting is less appropriate than the analytic evaluation of the physicist's theory or the engineer's design. The physicist does not complain that her theory will be analysed to determine whether it accurately models the collisions of the protons, because formulating theories that accurately model the universe is the concern of physics. Similarly, the engineer does not complain that his design will be analysed to determine whether it would collapse under high winds, because designing structures that fulfil such human requirements is what engineering is all about. Art, however, is not concerned with creating works that withstand the analytic evaluation of art critics. Art is about creating works that withstand the aesthetic evaluation of the beholder, be it the artist, the art critic, or the art gallery visitor. Analytic evaluation of a painting may reveal whether the brush strokes are broad or fine, whether the interplay between the different elements is strong or weak, whether the artist has been influenced by this or that art movement, and this can make a significant contribution to the beholder's appreciation of the painting. But, unlike the analytic evaluation of the physicist's theory or the engineer's design, the analytic evaluation of the artist's painting is of secondary importance in deciding the value of the creation. It is the aesthetic evaluation of the painting that is of primary importance.

Aesthetic evaluation is quite different from analytic evaluation. Analytic evaluation involves assessing the value of a creation according to definable criteria. When other scientists evaluate the physicist's theory, they analyse whether it accurately models the collisions of the protons. This criterion for the evaluation of the theory is well defined, there being little scope for disagreement as to whether or not the theory accurately models the collisions. Similarly, when art critics evaluate the artist's painting, they analyse, for example, how the artist has manipulated the brushes and paint. This criterion for the evaluation of the painting is also well defined, there being little scope for disagreement as to whether the brush strokes are broad or fine. In contrast to analytic evaluation, however, the aesthetic evaluation of a creation does not involve the protracted analysis of the creation according to definable criteria. As already suggested, it is impossible to define what makes a work of art good, so a work of art cannot be adequately evaluated by analysis according to definable criteria. Instead, the aesthetic evaluation of a creation involves a more immediate reaction to the creation.

Creativity for which the creation can be evaluated aesthetically is what I will call 'aesthetic creativity'. The artist exercises aesthetic creativity.

Just as analytic creativity is not confined to science and technology, so aesthetic creativity is not confined to art. Architects want their buildings to be beautiful as well as functional. Engineers, too, want their structures to be beautiful, though functional constraints in engineering often leave little scope for aesthetic creativity. Structures often derive their beauty from the economy with which the engineer has fulfilled the functional constraints, so that analytic rather than aesthetic creativity has contributed most to the beauty of the structure. For example, the colour of the paint on a suspension bridge is a purely aesthetic consideration, while the curves of the suspension cords is a purely analytic consideration, dictated by mechanics, yet the beauty of the bridge derives more from the curves of the suspension cords than from the colour of the paint.

Even science, which, more than any other discipline, might be expected to demand only analytic creativity, also demands aesthetic creativity. While the aim of a scientist is to formulate a theory that accurately models the universe, he may proceed by seeking to formulate a beautiful theory. The implicit assumption is that the universe is beautiful, so that theories that model it accurately must also be beautiful. This arbitrary assumption has proved a remarkably reliable catalyst for scientific discovery, not because the universe is objectively beautiful, but because only those who consider the universe to be beautiful – and so also consider theories that accurately model the universe to be beautiful – tend to become scientists. So scientists, too, indulge in aesthetic as well as analytic creativity.

Human creativity in the various fields of art, science and technology, then, tends to involve a combination of analytic and aesthetic creativity in different measures. In other words, most human creations can be evaluated both analytically and aesthetically. It may appear perverse to elaborate a distinction between analytic and aesthetic creativity, only to conclude with the claim that creativity tends to involve a combination of the two. However, the distinction is important as the processes of creativity are quite different for analytic and aesthetic creativity.

Exploration and Evaluation

Creativity proceeds through a combination of exploration and evaluation. Exploration is the process of generating ideas, and evaluation is the process of separating the good ideas from the bad ones. Exploration and evaluation do not occur as separate processes, representing different phases of the creative process, but are inseparable aspects of a single process.

Consider a composer writing a symphony. He explores different rhythms, tunes, phrases, melodies, themes, different combinations of notes, different combinations of instruments. He first explores and evaluates these different possibilities in his head. Those he considers to be no good, he rejects. He may then use a piano to hammer out the remaining possibilities, performing further exploration and evaluation. Again, the possibilities he considers to be poor, he rejects. He may then record the remaining possibilities on paper. Through exploration and evaluation of different combinations of these possibilities, the composer eventually works them into the final score.

Different composers work in different ways. Mozart is commonly credited with the ability to compose an entire work in his head before setting it down on paper. Historians believe this to be an exaggeration, but the myth probably contains an element of truth. Mozart probably performed more of his exploration and evaluation of musical ideas in his head than did other composers. Beethoven became increasingly deaf towards the end of his life, and was so desperate to hear his music as he composed it that he used to pound his pianos until the strings broke. Nonetheless, he continued to compose even after he became almost completely deaf, and even composed some of his most celebrated works at this time. So, towards the end of his life, Beethoven also probably performed more of his exploration and evaluation of musical ideas in his head. But regardless of whether a composer explores and evaluates ideas in his mind, on a piano or on paper, he always proceeds through a combination of exploration and evaluation.

Consider the physicist investigating collisions of protons in a particle accelerator. She performs her own experiments, or studies experiments performed by other physicists, to provide experimental data for her research. She explores ideas suggested by the experimental data, or ideas proposed by other physicists. She may formalise the ideas by expressing them in mathematical terms. She may analyse how accurately the ideas fit the observed data. She may also assess whether the ideas are consistent with current theories, may challenge established assumptions, may explore alternative theories. Again, the physicist first explores and evaluates ideas in her head. Those she considers to be no good, she rejects. She may then record the remaining ideas on paper or program them into a computer for further analysis. But again, regardless of whether the physicist explores and evaluates ideas in her head, on a computer or on paper, she always proceeds through a combination of exploration and evaluation.

The ideas that the composer and the physicist generate through exploration are not random. If creativity were a simple matter of generating random ideas through exploration then selecting the good ones through evaluation, it would take an eternity to create anything good. There are simply too many bad ideas for the creator to be so indiscriminate. Just as the monkeys typing at random would eventually come up with the complete works of Shakespeare but would waste eons typing nonsense, so the composer and the physicist might eventually come up with some good ideas, but would waste most of their lives exploring and evaluating bad ideas. Instead, exploration involves the generation not of random ideas but of generally good ideas. Most of the ideas won't be especially good, and some will be downright awful. But the composer's ideas are generally quite pleasing to the ear, and the physicist's ideas generally represent quite feasible models of the collisions of the protons. After subsequent evaluation, the composer may decide that most of his ideas are not good enough for inclusion in his symphony, and the physicist may decide that most of her ideas are inconsistent with the experimental data. But even these rejected ideas may represent the seeds of further ideas.

Humans can explore ideas in an unlimited variety of ways, so no account of the process of exploration can be definitive. Nonetheless, brief descriptions of some common ways in which we explore ideas not only give a flavour of the powerful simplicity of creative thinking, but also explain how a creator is able to generate not random ideas, but generally good ones.

Consider the physicist investigating a particular collision of protons in a particle accelerator, one that is not explained by current theories. She will spend much of her time thinking not about the collision itself, but about the particles and forces involved in the collision, and how these relate to current theories. As she considers these particles, forces and theories, she will recall other concepts through association, some directly related to the collision, some apparently unrelated. She may analyse patterns in the properties of the particles and the forces, as well as the assumptions and deficiencies of current theories. She may play with the physical concepts involved in the collision – mass, momentum, position, probability – and combine and transform these concepts in unexpected ways. She may devise visual representations of the collision, and explore the insights that different representations may yield. The collision may remind her of analogous phenomena, either in physics or in other fields, and she may study such analogies further.

Consider the composer writing a symphony. He may analyse the constraints imposed by the current conventions for symphonies, and investigate the possibilities afforded by the removal of these constraints. For example, he may consider using notes that are not part of a conventional scale, rhythms that do not fit a conventional time signature, or instruments that are not part of a conventional orchestra. He may impose his own constraints. Within whatever constraints he chooses, he will explore different combinations of sounds, recognising patterns in the combinations and relationships between the combinations. In this way he will subconsciously accumulate an enormous repertoire of sounds, cross-referenced through the associations between the entities that represent them in his mind. He will associate the sounds not only with each other, but also with the feelings they evoke, and when he wants to evoke a certain mood in the symphony, he will recall the appropriate sounds through association. He will combine and transform sounds to form new sounds to add to his repertoire, and he may devise visual representations of sounds, and manipulate these visual representations to explore the sounds.

These descriptions of exploration may seem mundane, far removed from the popular notion that creativity is an inspired activity. Nonetheless, ordinary or extraordinary, these are the ways in which exploration proceeds: association, recognition of patterns, recognition of relationships, recognition of analogies, analysis of concepts, combination of concepts, transformation of concepts, analysis of constraints, removal of constraints, imposition of constraints, invention of representations. This list is not comprehensive. It could not be comprehensive. Creative thinkers always invent new ways to explore ideas.

The power of these ways of exploring ideas lies in their flexibility. They are not limited to the manipulation of concepts in

physics and music, but they can be used to manipulate concepts in any field. They do not limit creative thinking to the
realms of the logical, or the feasible, or the reasonable, but comfortably accommodate the realms of the fantastic and the
impossible and allow the thinker to flit between ideas that are
completely unrelated in any analytic sense. This flexibility
derives from the associative nature of our thought, as described
in the previous chapter. If human thought were based on logic
rather than association, if every concept and every relationship
we considered were defined precisely rather than loosely
through associations, then we could never be creative.

It can now be seen how the process of exploration generates
not random ideas, but generally good ideas. Consider the physicist investigating the collision of protons. The concepts that she
explores are not random, but are connected with the collision of
protons in some way. The connections might be tenuous. For
example, the pattern of the particles created in the collision
might remind her of the pattern of the pepperoni slices on the
pizza she ate last night, or the classification of particles by physicists might remind her of the classification of wines by vintners.
But some connection, tenuous or substantial, always exists, so
that the ideas the physicist generates are not random, but generally relevant to the collision. Further, the physicist explores
these concepts not randomly, but in ways that she knows to be
productive. She may use standard techniques that physicists of
the past have found to be productive. She may also use novel
techniques that she has invented herself specifically for exploring the collision. Her experience in exploring physical concepts,
combined with the collective experience of other physicists,
ensures that the ideas she generates are not random, but generally good.

Consider the composer writing the symphony. He too generates not random ideas, but generally good ideas. As he explores
sounds, he will select only those sounds that he considers to be
good for further exploration. As discussed in the previous chapter, this more elaborate processing will decorate the entities

representing the good sounds with more associations, so that the good sounds will be considerably easier to recall than the rejected sounds. The repertoire of sounds the composer accumulates, then, will not be of random sounds, but of good sounds. Further, the composer explores these sounds not in random ways, but in ways that he knows to be productive. He may have studied the techniques of composers of the past, and so have the benefit of centuries of experience. He may also have developed techniques of his own. His experience in exploring sounds, combined with the collective experience of the composers of the past, ensures that the ideas he generates are not random, but generally good.

These factors, while improving the chances that the physicist's and the composer's ideas will be good, do not guarantee that their ideas will be good. The physicist's association of the pattern of the particles created in the collision with the pattern of the pepperoni slices on a pizza is unlikely to yield any great insights into particle physics. This makes the process of exploration something of a hit-and-miss affair. Sometimes, someone exploring ideas in one field may come up with an idea in an entirely different field. So the physicist may fail to formulate a theory that explains the collision of protons, but she may come up with a great concept for a new pizza.

The process of exploration is much the same whether the creativity is analytic or aesthetic. Different techniques are used in different fields. The physicist's most useful technique may be the scientific analysis of concepts, whereas the composer's most useful technique may be the combination and transformation of sounds. These different techniques demand different knowledge and different skills, but the basis of the process of exploration, which is the manipulation of concepts, is the same in all fields. In contrast, the process of evaluation is fundamentally different for analytic and aesthetic creativity.

For analytic creativity, the ideas generated through exploration are evaluated through analysis according to definable criteria. For the engineer, the process of evaluation involves analysing his design to determine whether the bridge would carry the required load over the required distance with the required reliability. This involves complex mathematical techniques, and, increasingly, complex computer models. For the physicist, the process of evaluation involves analysing her theory to determine whether its predictions are consistent with observations of collisions of protons, how it relates to other theories, and so on. Again, this analysis often involves complex mathematical techniques and complex computer models. The engineer and the physicist require expert knowledge and skills in their respective fields to be able to analyse ideas in this way.

For aesthetic creativity, the evaluation of the ideas generated through exploration is not so easily explained. Without definable criteria according to which an idea can be analysed, evaluation is not a rational or intellectual exercise, but a more basic process. Aesthetic evaluation tends to be immediate. The composer knows immediately that the phrase he has just come up with is good for the particular part of the symphony he is composing. The physicist knows immediately that the beautiful hypothesis she has just formulated is good, before she has a chance to evaluate it analytically. Aesthetic evaluation also tends to be certain. The creator *knows* whether the idea is good. This seemingly supernatural knowledge demands explanation.

Our aesthetic evaluation of an idea is an artefact of our emotional response to it. Whether we consider an idea to be good or bad is determined entirely by our emotions.

Intellect has a role in the creation of a work of art. An artist's intellectual analysis of her ideas may allow her to develop these ideas and may prompt further ones. For example, she may think of using broad brush strokes to represent the wires between the telegraph poles in her painting, even though, in reality, the wires are fine. She may analyse the idea intellectually, considering how it will affect the balance of the painting and how the beholder of the painting will interpret the exaggeration of the wires. Her analysis may suggest a development of the idea, such as the representation of all the artificial features of the painting in broad brush strokes to contrast with the more accurate representation of the natural features. Her analysis may prompt further ideas, such as using different palettes to represent artificial and natural features.

However, the artist's intellectual analysis of the idea of using broad brush strokes to represent the wires does not determine whether she considers it to be a good idea. Instead, it is her emotional response to the idea that determines whether she considers it good. Perhaps the exaggeration of the wires elicits a feeling of anger associated with the encroachment of the trappings of technology on the natural environment. If this is a feeling that the artist wants to communicate, she may consider the idea to be good. In art, analytic evaluation, which involves intellect, is of secondary importance to aesthetic evaluation, which involves emotion.

Emotion is more basic than intellect. Analytic evaluation tends to be protracted, whereas aesthetic evaluation can be immediate. Further, analytic evaluation tends to demand the attention of the creator, whereas aesthetic evaluation can be subconscious. It is because aesthetic evaluation can be immediate and subconscious that the processes of exploration and evaluation are inseparable. When an idea is generated subconsciously, it can be immediately and subconsciously subjected to aesthetic evaluation. Good ideas can be revealed to the conscious mind for further exploration and evaluation. Bad ideas can be rejected without ever being revealed to the conscious mind. This contributes further to the creator's apparent ability to generate only good ideas. Not only does the process of exploration tend to generate generally good ideas, but the process of aesthetic evaluation can filter out bad ideas without the creator's ever being aware of them.

The primacy of emotion over intellect in the evaluation of a work of art explains why it is difficult to describe what makes it good. Aesthetic evaluation, which involves emotion, precedes the verbalisation and rationalisation of that evaluation, which involves intellect. The primacy of emotion over intellect also explains why it is impossible to define what makes a work or an idea good. Aesthetic evaluation, which involves emotion, precedes the analysis and definition of that evaluation, which involves intellect. Verbalisation, rationalisation, analysis and definition of what makes a work of art good are secondary impositions on the primary emotional response.

Creativity, then, is a combination of exploration and evaluation. The creator generates ideas through the manipulation of concepts, and separates good ideas from bad ones through intellectual analysis and through emotional responses to them.

The Mystery of Creativity

Creativity possesses an air of mystery. If the less-than-mysterious conclusion of the previous section is to be accepted, then this air of mystery must be dispelled.

The inaccessibility of many creative works contributes to the perceived mystery of creativity. Einstein's theory of relativity is extolled by physicists, but people who lack knowledge of mathematics are often intimidated by the complex symbols and equations used to describe the theory. Picasso's paintings are admired by historians of art, but people who lack confidence in their own aesthetic evaluation of his paintings are often intimidated by art historians' apparently definitive appraisals of them.

It is revealing that attempts to make science and art more accessible can be extremely successful. Standard textbooks on Einstein's theory of relativity describe the concepts required to understand the theory in mathematical terms, using the language of covariant derivatives, geodesic deviations and stress-energy tensors. (Einstein himself once quipped: 'Since the mathematicians have attacked the relativity theory, I myself no longer understand it any more.') Popular books describe the same concepts in more familiar terms. They use the language of space and time and curvature, and explain in plain English how our concepts of space and time must be revised to describe the universe on a large scale. In place of the symbols and equations are pictures showing how objects floating weightless in a spaceship move towards each other as the spaceship falls towards the earth, how light from distant stars is deflected as it passes close to the sun, how an astronaut falling into a black hole appears elongated to an outside observer. Without the mathematics, the reader of a popular book will not be able to perform the complex calculations required to predict the behaviour of the universe in all circumstances, but from the specific instances illustrated in the popular book, he will gain a deep understanding of the nature of the universe. The symbols and equations in standard textbooks will then seem not so much mysterious as superfluous.

The language of art, too, can be intimidating. Appraisals of Picasso's paintings may use such off-putting terms as 'illusionistic space', 'stereometric form', 'analytical cubism'. However, it is not only the language used to describe art that is inaccessible: the art itself can be inaccessible too. Picasso's paintings do not represent reality in the same way as a photograph does. In some of his paintings, he divided the canvas into geometric shapes, each depicting a different part of an object such as a violin from a different direction. In others, he simplified human faces to line drawings, their features displayed in unnatural positions and orientations. To some people, these paintings are unattractive, and Picasso's ways of representing reality perverse.

Attending an art appreciation course might change these opinions. Such a course might describe in plain English the concepts behind the inaccessible terms mentioned above, but this would be the least of its contributions to the student's appreciation of Picasso's paintings. More important is the understanding that no painting can represent reality as it really is, only as the artist sees it, and that, moreover, part of the value of painting lies in its allowing others to see reality as the artist sees it. Picasso's line drawings of faces with their features awry may not be true to the anatomy of the human face as it is described in medical textbooks, but it is true to reality as Picasso saw it. Indeed, human emotions are perhaps more dramatically portrayed in these drawings than they could ever be-in more conventional paintings. The more the student studies art, the more she may come to value her own feelings about a painting. Like the symbols and equations in the standard textbooks on relativity, the terms used by art historians will then seem not so much mysterious as superfluous.

The idea that only brilliant scientists and artists can be creative also contributes to the perceived mystery of creativity. I have described creativity as requiring that the creation go beyond the previous experience of the creator, but creativity is commonly considered to require instead that the creation precede all similar creations in time. So science is considered creative only if the scientist formulates a paradigm radically different from the currently accepted paradigm, and art is considered creative only if the artist develops a style radically different from all previous styles. Since only the most brilliant scientists succeed in formulating a radically different paradigm, and only the most brilliant artists succeed in developing a radically different style, a common conclusion is that most people are incapable of creativity. This misconception is self-reinforcing. It discourages people who do not consider themselves brilliant from indulging in creative activities, and so forever forestalls their discovery that they are indeed capable of creativity.

Even people who reject the idea that they are incapable of creativity, and so indulge in a creative activity such as painting, are often discouraged by their slow progress. It can take many hours of practice for a beginner to acquire the technical skills required to produce a good painting. Many novice painters regard their initial technical incompetence as evidence of a lack of 'natural talent', and give up quickly. Others lack the time, resources or will to persevere in their efforts, and again soon abandon their new pastime. Thinking that their failure betrays an all-encompassing inability to be creative, these failed novices are often reconfirmed in the belief that creativity is a mysterious talent that most people lack.

The notion that creative ideas are conceived in a flash of inspiration further contributes to the air of mystery. The popular perception of the creative process is of the creator considering and rejecting one idea after another, until, suddenly, in a flash of inspiration, he conceives a brilliant idea from nothing. The flash of inspiration is generally accompanied by an exclamation of 'Eureka!' or some such mystical intonation. The notion that brilliant ideas can come from nothing is bolstered by the tendency of artists and scientists to publish only the end product of their creativity, a complete work or theory. Evidence of the hard work that made the creation possible, such as the artist's sketchbook or the scientist's notebook, tends to come to light only years later, and even then only scholars take note of it. Such evidence, of course, supports the more mundane description of the creative process proposed above, according to which brilliant ideas do not come from nothing, but from the creator's manipulation of concepts.

The notion that brilliant ideas can be generated from nothing might be rejected without further discussion, if it were not for the claims of respected scientists and mathematicians to have had flashes of inspiration. While countless scientists and mathematicians have had sudden insights, only a handful have claimed these to be anything other than the culmination of prior creative thinking. So there exist just a few, oft-repeated accounts of brilliant ideas that have seemed to come from nothing. It should also be stressed that scientists' and mathematicians' reports of their own sudden insights should be approached with scepticism, because, as the history of psychology shows, introspection is a notoriously unreliable psychological tool. Introspective reports often reveal more about the reporter's expectations of how the brain should work than about how the brain really works, and this applies as much to the introspective reports of respected scientists and mathematicians as to those of anyone else. Nonetheless, the claims of flashes of inspiration can hardly be ignored. Even if the scientists' and mathematicians' perception that their ideas come from nothing is illusory, an explanation for this perception is demanded.

Creativity, for the most part, is hard work. Without exception, the greatest artists and scientists throughout history have dedicated years of their lives to acquiring knowledge of their particular fields, techniques for the manipulation of that knowledge, and skills for the practice of their art or science. Thus equipped, they have dedicated further years to developing and exercising their creativity. Mozart started to learn music, according to a strict régime imposed by his composer father, at the age of four. He started to compose music, again with considerable help from his father, at the age of six. So by the time he started to compose his mature works, he was able to draw on an intimate knowledge of music, well developed skills for playing musical instruments, and years of experience of composition. (Idolisers of Mozart would like to think that even his childhood compositions were consummate, but the truth is that they were no more than extremely good for his age.) Similarly, Einstein declared at the age of twelve that his vocation was to unravel the mysteries of the universe. He spent four years studying physics at university followed by another five after graduation before he published his first papers, including the special theory of relativity. It was another eleven years after the publication of the special theory of relativity before the publication of the general theory of relativity, a reformulation of the special theory to encompass gravity and acceleration. Einstein did not spend these years patiently waiting to be inspired, but working hard on the problem. Both Mozart and Einstein are considered geniuses, but neither achieved his creative works without years of hard work.

A prolonged period of hard work generally precedes the flashes of inspiration claimed by scientists and mathematicians. This is highly suggestive that their brilliant ideas do not come from nothing, but represent the fruits of that hard work. Sometimes, though, the flash of inspiration is reported as occurring during a break from the hard work, when the creator let his mind wander to other things. Archimedes is said to have thought of the idea of measuring volume by displacement of water while in the bath; Kekulé thought of the idea that the benzene molecule might have a circular structure immediately after waking from a light sleep; Poincaré thought of an important property of the set of mathematical functions he called 'Fuchsian functions' while stepping on to a bus at the start of an excursion. In each of these cases, though, the scientist or mathematician had been working hard on the problem for considerable time before the break that augured the breakthrough.

Scientists and mathematicians tend to think analytically. As suggested above, the process of exploration may be most flexible, and most productive, when the thinker flits between ideas which seem completely unrelated in any analytic sense. Scientists and mathematicians are particularly susceptible, then, to the creative trap psychologists call fixation, whereby they are unable to break out of fixed patterns of thought, unable to stop thinking analytically and to start thinking by association. Fixation can often be overcome simply by taking a break from thinking about the problem in hand and letting the mind wander to other matters. (Anyone whose work requires analytic thinking knows how valuable frequent coffee breaks can be for overcoming fixation. Unfortunately, their supervisors may be less convinced of the value of such breaks.) It seems likely, then, that Archimedes' bath, Kekulé's nap and Poincaré's excursion served merely to overcome fixation.

The very moments of inspiration of Archimedes, Kekulé and Poincaré can be explained in terms of subconscious exploration and evaluation. Each of the creators had been thinking about his respective problem intensively in the period prior to his sudden insight. So the relevant concepts were at the back of his mind, or, to use the terminology of the associative model of memory, the entities representing these concepts were highly activated. During his break from thinking about the problem, the creator's attention was unfocused: Archimedes was relaxing in the bath, Kekulé was waking from a nap, and Poincaré was stepping on to a bus. The subconscious mind was free to explore the concepts so recently set aside by the conscious mind. Whereas the conscious mind is capable of complex analytic thinking, the subconscious mind is capable only of simple associative thinking, so while the creator's conscious exploration may have been hindered by inflexible analytic thinking, his subconscious exploration was able to discover relationships between concepts through association that would never have been suggested by analysis.

This subconscious exploration would have no immediate benefit to the creator, being unavailable for conscious evaluation, were it not for the possibility of subconscious aesthetic evaluation. Remember, a scientist's initial evaluation of an idea is generally aesthetic. The scientist knows instantly that the beautiful hypothesis he has just formulated is good, before he has a chance to evaluate it analytically. The same is true for mathematicians. Subconscious exploration, then, can be coupled with subconscious aesthetic evaluation of the ideas so generated. As Archimedes', Kekulé's and Poincaré's subconscious minds explored relationships between concepts through the associations between them, each newly discovered relationship prompted an emotional response. For the most part, these emotional responses were neutral or negative, as the relationships were trivial or worthless. Eventually, though, the associations formed through the prolonged period of hard work prior to the break revealed a relationship that was neither worthless nor trivial, a relationship that fitted simply and beautifully with the concepts over which the creator had previously deliberated. This relationship prompted an emotional response that was neither negative nor neutral, but overwhelmingly positive. Any creator will know the surge of feeling that accompanies the generation of a good, or beautiful, or revolutionary idea. It is this positive emotional response that calls the attention of the creator to the idea, and prompts the revelation of the idea to his conscious mind.

No wonder, then, that such sudden insights should seem to come from nothing, should seem inspired. The creator is not consciously thinking about the problem at the time he has the idea, so it is not surprising that the idea should seem to come from nothing. The positive emotional response precedes the revelation of the idea to the conscious mind, indeed, prompts this revelation, so it is not surprising that the idea should seem inspired from without rather than generated from within. Both the exploration and the evaluation are subconscious, so it is not surprising that introspective reports should give no explicit account of these processes.

Further, it is not unexpected that scientists and mathematicians, particularly good scientists and mathematicians, should be susceptible to sudden insights, because the best ideas in science and mathematics are the simplest. Einstein's first step towards his formulation of the general theory of relativity was his realisation that a person falling freely will not feel his own weight. (This may seem obvious to the reader familiar with the concept of weightlessness, but it was not obvious in the days before air flights were common and space flights were possible.) Einstein described this realisation as a sudden insight (though he did not describe it as coming from nothing). However, the idea that a person falling freely will not feel his own weight is so simple that it is impossible to imagine its genesis other than as a sudden insight. Einstein's previous reflections on gravity and acceleration had prepared the ground for this idea thoroughly, so that when it occurred to him, he immediately recognised its import. But the idea itself could only have come all at once. Einstein could not have had one half of the idea one moment, and the other half of the idea some moments later, because the idea that a person falling freely will not feel his own weight is a single, irreducible idea. I do not want to denigrate Einstein's achievement: the simplicity of his idea did not make it any less revolutionary, indeed, made it all the more revolutionary. All I want to suggest is that Einstein could only have had the idea as a sudden insight, could only have realised it in a single moment.

Little by little, the air of mystery about creativity is dispelled. The only mysteries that remain are those of the emotional response that is the basis of aesthetic creativity, and those of the subconscious exploration and evaluation that can be so important to both analytic and aesthetic creativity. Emotional responses seem less mysterious, because we experience them constantly. Subconscious exploration and evaluation seem more mysterious, because, by definition, we do not experience subconscious processes, so we have more difficulty in accepting their existence. These last remaining mysteries are dispelled in the following chapters on emotions and on consciousness.

It is sometimes suggested that when the mysteries of a phenomenon such as creativity are dispelled, something valuable is lost, something that can never be regained. While it is true of all attempts to expand our knowledge that innocence is lost, the value of that innocence is questionable. In the case of an attempt to understand creativity, I consider that what is gained from such an understanding vastly outweighs what is lost.

Lost is our idolisation, our near-deification, of brilliant creators such as Mozart and Picasso, Archimedes and Einstein. Lost is the perception of an unbridgeable gulf between these brilliant creators and mere mortals. Lost is the conviction that most people can never hope to do anything that remotely compares with the achievements of these brilliant creators.

What is gained from an understanding of creativity is not just the satisfaction of knowledge for its own sake, but also the liberating realisation that creativity is not the preserve of an élite of brilliant minds, but an ability common to all humans. Exploration requires specialised knowledge and skills, but these can be learnt. Techniques for exploration, such as techniques for the recognition of patterns, relationships and analogies, can also be acquired. Like exploration, analytic evaluation requires specialised knowledge and skills, and again, these can be developed. Aesthetic evaluation requires a willingness to recognise emotional responses to ideas, emotional responses that need not be learnt, indeed, cannot be learnt. Otherwise, creativity requires only motivation and commitment.

Not everyone is able to master the skills required to paint in the exacting style of the artists of the High Renaissance. Not everyone can acquire the knowledge required for a thorough understanding of Einstein's theory of relativity. But most people are able to learn the skills required to play a tune on a piano, or draw a sketch, or write a short story, or investigate mathematical patterns on a computer. The perceived mystery of creativity, particularly people's perception that there is some objective measure of what is good to which they cannot hope to aspire, tends to discourage people from developing their knowledge and skills, and so from giving greater rein to their creativity. It is notable that psychologists attempting to understand creativity find that the volunteers who participate in their experiments are generally surprised at their own capacity for creativity. Indeed, the volunteers often find the creative activities they perform in the laboratory so satisfying that they are enthusiastic to continue them at home. If the demystification of creativity allows people to reject the notion that they are incapable of being creative, and so encourages them to seek the satisfaction that such activities can afford, then something of value is surely gained.

The Creative Computer

For human creativity to be reproduced on a computer, it must be capable of exploration, and of both analytic and aesthetic evaluation.

The different ways in which humans are able to explore ideas derive from various aspects of human intelligence. For example, our ability to explore relationships between concepts through association derives from the associative nature of our meaning-based memory. Our ability to combine and transform sounds and images in our heads comes from the availability of our perception-based memory for mental manipulations. (Our devising visual representations of concepts increases the possibilities for exploration, because concepts encoded both as entities in our meaning-based memories and as images in our perception- based memories can be manipulated both through association and through combination and transformation of images.) Our recognition of patterns and analogies derives from the same processes as our recognition of objects for visual perception. The repertoires of possibilities we accumulate through exploration (such as the repertoire of sounds the composer accumulates) are learnt in the same way as all our knowledge. The techniques we use to manipulate concepts (such as the application of logic) are developed in the same way as all our skills.

We have established that each of these aspects of human intelligence – our meaning-based memory, our perception-based memory, our recognition of objects, our learning of knowledge, our learning of skills – might be reproduced on a computer. Because the various ways in which humans are able to explore ideas derive from the various aspects of human intelligence, a computer on which human intelligence were reproduced would be able to explore ideas in the same ways. Exploration is no more than the use of intelligence for the generation of ideas. The intelligent computer, like the intelligent human, is capable of exploration.

The same is true of analytic evaluation. The knowledge and skills humans use for the analytic evaluation of ideas (such as the engineer's ability to devise computer models of his bridge and the physicist's development of computer models of her theory) are learnt in the same way as all our knowledge and skills. Again, we have established that human learning of knowledge and skills might be reproduced on a computer. Analytic evaluation is no more than the use of intelligence for the analysis of ideas. The intelligent computer, like the intelligent human, is capable of analytic evaluation.

Exploration and analytic evaluation, then, could be reproduced on a computer by making it intelligent. Aesthetic evaluation is a different matter. The aesthetic evaluation of an idea requires not intelligence, but an emotional response to the idea. If a computer is to be capable of aesthetic evaluation, it must be capable of emotion. The following chapter argues that computers could be emotional in the same way as humans, describing how human emotions might be reproduced on a computer, so that computers, like humans, might then be capable of aesthetic evaluation.

This discussion of creativity, like the earlier discussion of intelligence, has studiously avoided two important questions: the question of motivation, of what induces us to explore and evaluate ideas; and the question of will, of the nature of our decisions to explore or not to explore ideas, and to pursue one idea in preference to another. These questions are discussed in the following chapters on emotions and on consciousness, but it is worth emphasising the particular importance of motivation to creativity. It is no accident that the greatest creators in history have all been highly motivated, committed to and deeply involved in their work. Because creative activities hold no guarantee of results, motivation can be a problem for the creator. It can be difficult for the artist to motivate herself to explore ideas with no guarantee that her exploration will yield a work of any value. It can be difficult for the scientist to motivate himself to explore concepts without the certainty that his exploration will arrive at an understanding of the phenomenon he is investigating. If, however, the creator's efforts do produce results, the pleasure of creating something beyond his previous experience, something he considers to be good, can be extremely rewarding. Even if the creator's efforts are not successful, the process of exploration can be as much fun for the would-be creator as play is for a child. Indeed, the satisfaction the creator derives from manipulating ideas in original ways is similar to the satisfaction the child derives from manipulating toys in original ways.

I have proposed that for a computer to be described as creative, it must create something beyond its previous experience, something it considers to be good. Before a computer could begin to be creative, then, it would have to acquire its own experience of the world and develop its own notion of what is good. It is unlikely that Beethoven could have composed his Ninth Symphony in his deafness if he had not previously heard the sounds of a chorus of human voices and an orchestra of instruments. Einstein could not have formulated his theory of relativity if he had no feeling for the simple beauty of the universe. Wordsworth could not have written that he wandered 'lonely as a cloud' if he had never been lonely, or never seen a cloud drift slowly across the sky. It would be absurd to expect a computer to be similarly creative if it had not led a similarly rich existence. Once the computer had acquired its own experience of the world and developed its own notion of what is good, it could begin to be creative. I have proposed that creativity demands intelligence for the exploration and analytic evaluation of ideas, and emotion for the aesthetic evaluation of ideas. Creativity, then, could not be reproduced on a computer independently of intelligence and emotion. The creative computer will not be realised through half-measures.

Human creativity is not a mysterious ability exclusive to an élite of brilliant minds, but a universal ability that arises from our intelligence and our emotion. If human intelligence and human emotion can be reproduced on a computer, then the computer could be creative in the same way as we are.

3

THE EMOTIONAL
COMPUTER

Emotions

Humans feel countless emotions. Some are associated with simple sensory stimuli: hunger with an empty stomach, thirst with a dry throat, pain with an injury. Some are associated with more complex sensory stimuli: fear with danger, embarrassment with a blunder, love with a loved one. Others are associated with sensory stimuli only indirectly: sorrow, hope, disappointment. Others still are not described by a single word in the English language: wanting to be the centre of attention, desiring admiration, wanting to excel.

We often think of our emotions as what makes us human. Certainly, someone who fails to exhibit the emotions of sympathy and compassion may be called inhuman. People have long been reluctant to concede that other animals may have emotions in the same way as humans, yet it is only by regarding humans in the same way as other animals, as a product of billions of years of evolution, that the origin of emotions can be understood.

Life on earth began with molecules that were capable of replication. From the resources available in the seas of the young earth, nearly four billion years ago, these replicators made exact copies of themselves through simple chemical reactions. Some made better use of the resources than others, and so replicated more effectively, and so survived. Others replicated less effectively, and perished. Sometimes the process was imperfect, so that the child replicator differed in some respect from its parent. Where the difference was such that the child made better use of resources, and so replicated more effectively, it would flourish at the expense of its cousins. This natural variation precipitated the development of replicators into ever more complex organisms. Modern evolutionary theory is based on one simple principle: those replicators that replicate survive, those that don't perish.

An important development in the history of life on earth occurred when the replicators began to build survival machines that allowed better use of available resources, and thus more effective replication. The machine-building replicators flourished at the expense of their cousins. As before, natural variation allowed the evolution of ever more complex machines. The machines came to dwarf the replicators, so much so that the first proponents of evolutionary theories, unaware of the importance of the microscopic replicators, based their theories around the macroscopic machines. Modern evolutionary theory is based instead around the replicators. Our genes are the replicators, complex DNA molecules descended from the simple molecules with which life began, and our bodies are the survival machines, constructed according to the instructions encoded in our genes to effect their replication.

Another important development in the history of life on earth occurred when the replicators began to build machines with central nervous systems, in other words, when animals began to develop brains. The primary evolutionary advantage of a brain was its ability to control the movement of the animal by mediating between ever more complex sensory and motor mechanisms.

In an animal with simple sensory and motor mechanisms, such mediation is unnecessary. The microscopic amoeba moves through water by changing the shape of its single-cell body, extending part of it in the direction of motion and following through with the rest of its body (see Figure 19). This movement is controlled by chemical signals. Chemicals favourable to the amoeba's survival prompt the extension of the part of the amoeba's body in which they are detected. For example, chemicals emitted by food prompt the part of the amoeba's body closest to the food to extend, so that it moves towards the food. Chemicals detrimental to the amoeba's survival prompt the opposite response in the part of its body where the chemicals are detected, so the amoeba moves away from water that is too acidic or too alkaline. The amoeba's responses are local, in that a sensory stimulus and the motor response it prompts occur in the same part of the amoeba's body. Further, the amoeba's responses are immediate, in that the motor response follows the sensory stimulus without delay. So no central nervous system is required to mediate between the sensory and motor mechanisms.

The amoeba's simple sensory and motor mechanisms are sufficiently simple that any part of its body can detect the chemicals that are beneficial or detrimental to its survival, and be extended to initiate motion. However, as soon as animals developed more complex sensory and motor mechanisms, this symmetry was unsustainable. A complex sense organ such as an eye is so expensive to grow that it would be evolutionary suicide for an animal to grow eyes all over its body. If, instead, the

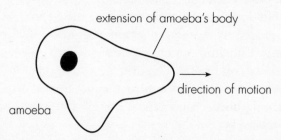

Figure 19. An amoeba moves through water by changing the shape of its single-cell body, extending part of it in the direction of motion and following through with the rest of its body.

animal's genes dictate that only two eyes should be grown at strategic positions on the body, and that the resources that would have grown the remaining eyes should be allotted to other survival mechanisms, then the genes' chances of replication are considerably improved. Similarly, a complex motor mechanism such as a fish's tail is too expensive to be grown all over the fish's body, even though it would allow the fish, like the amoeba, to move in any direction without turning around. Instead, the fish's genes dictate that only one tail should be grown, and again the genes' chances of replication are considerably improved.

The development of discrete sensory and motor mechanisms required the simultaneous development of a means of communication between the two. Suppose the ancestors of the amoeba developed a whip-like tail for propulsion (some present-day relations of the amoeba possess just such a tail). This would be useless without some communication between the chemical-detecting parts of the tailed amoeba's body and the tail itself, so that the tail could propel the amoeba towards helpful chemicals and away from harmful ones.

Such an elementary nervous system would require primitive processing of sensory information. If harmful chemicals were detected in the part of the tailed amoeba's body closest to the tail, signals would be sent to the tail to indicate that it should thrust to propel the tailed amoeba away from the chemicals. If, instead, harmful chemicals were detected in the part of the tailed amoeba's body furthest from the tail, signals would be sent to the tail to indicate that it should spin the amoeba around. This would ensure that the tailed amoeba is pointing away from the chemicals before the tail thrusts to propel it forwards (see Figure 20 on page 148).

As animals developed increasingly complex sensory and motor mechanisms, the nervous system required to mediate between them became increasingly extensive. A sense organ as complex as an eye provides sensory information that requires complex processing if it is to be useful in directing the animal's movements. A motor mechanism as complex as a fish's tail

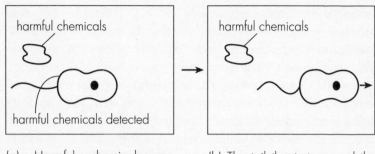

(a) Harmful chemicals are detected in the part of the tailed amoeba's body closest to the tail.

(b) The tail thrusts to propel the tailed amoeba away from the chemicals.

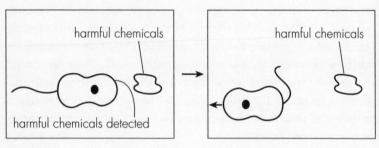

(c) Harmful chemicals are detected in the part of the tailed amoeba's body furthest from the tail.

(d) The tail thrusts to spin the tailed amoeba around before thrusting to propel it away from the chemicals.

Figure 20.

requires that the signals sent to the different muscles in the tail be precisely coordinated if the muscles are to contract and relax in concord to propel the fish forwards. So not only did the nervous system mediating between the sensory and motor mechanisms become more extensive, it also developed centres for the complex processing of sensory and motor information. Animals developed brains.

This development represented an important transfer of direct control over an animal's behaviour from its genes to its brain. The behaviour of an animal without a brain, such as an amoeba, is determined directly by its genes, which determine the structure of the animal's body, which in turn determines the behaviour of the animal. For all its directness, this mechanism is highly inflexible. An amoeba's genes can do no more than encode a limited number of responses to a limited number of circumstances: move towards food, move away from acids and alkalis, reproduce in favourable conditions, excrete a protective membrane in unfavourable conditions, and so on. The behaviour of an animal with a brain, such as a fish, is far more flexible. The animal's genes directly determine the structure of its body, including its brain, and provide general directions as to the behaviour required of the animal. Control over everyday decisions is then abdicated to the animal's brain. It is the fish's brain that processes the sensory information from its eyes to determine whether it is seeing a potential predator or prey, a potential competitor or mate. It is the fish's brain that controls the response to this information, processing the motor signals sent to the muscles in the tail and fins to propel the fish as appropriate. The genes' contribution is to provide the general directions as to the behaviour required: flee predators, eat prey, chase away competitors, reproduce with mates.

When genes build a survival machine, they cannot foresee every possible circumstance to which the machine may have to respond and encode the required response for each of these circumstances. There are simply too many possible circumstances. By providing the survival machine with a central processor of information capable of responding to unforeseen circumstances in an intelligent way, the genes achieve considerable behavioural flexibility at the cost of the abdication of direct control over the behaviour of the machine. As long as the brain follows the general directions provided by the genes, the abdication of control does not reduce the genes' chances of replication. The improved behavioural flexibility, though, considerably improves the genes' chances of replication.

As animals' brains developed – from simple mediators between sensory and motor mechanisms, to increasingly complex processors of information – the animals became increasingly intelligent, and more and more capable of deciding how to behave. This development hints at the intriguing possibility that an animal might decide not to dedicate its entire existence to the replication of its genes, but to act in ways that reduce, rather than improve, its genes' chances of replication. Having abdicated direct control of the animal's behaviour to the brain, the genes would be powerless to stop the animal's rebellion.

The rebellious animal might live a long and fulfilling life, doing its own bidding rather than following the general directions provided by its genes. But genes that allow their survival machine to rebel in this way would soon become extinct, their chances of replication reduced compared to those of genes that keep tighter rein over their survival machines. The lesson for genes that endow their survival machines with brains is clear. If they are to survive as a species, they must provide their survival machine with general directions so overpowering for the machine that it cannot ignore them, and cannot rebel against its genes.

Animals developed emotions to meet these requirements. It is through emotions that an animal's genes provide it with general directions as to how to behave. It is through emotions that the animal's genes maintain control over the animal's brain.

Consider a rabbit that suddenly becomes aware of a fox a short distance away. The fox is a direct threat to the rabbit's life. It is also a threat to the chances of replication of the rabbit's genes: a dead rabbit does not reproduce. The rabbit is overpowered by fear, and flees. Although it is fear that motivates the flight, it is the rabbit's brain that controls the flight, that coordinates the movement of the muscles in the rabbit's limbs to achieve motion, that remembers the direction of the closest hole into the warren.

Now consider a bear that suddenly becomes aware that her young cubs have strayed too close to some humans. The humans are a potential threat to the cubs' lives, but not to their mother's life, since she is a safe distance away. To the mother herself, rescuing her cubs might seem like a bad move, since it will put her own life at risk.

From her point of view, this behaviour would be called altruistic. From the point of view of her genes, though, the equation is different. The cubs inherit half their genes from their mother and half from their father, so each of the mother's genes has a fifty-fifty chance of being inherited by each of the cubs. The humans, then, are a threat not only to the cubs' lives, but also to the chances of replication of a good proportion of the mother's genes: dead cubs do not go on to reproduce. So for her genes, providing the bear with general directions that motivate her to retrieve her cubs is far from altruistic, it is simply a sound survival strategy. And it is the bear's genes that call the shots. The bear is overpowered by love for her cubs and vicarious fear for their lives, and although, again, love motivates the rescue, it is the mother bear's brain that masterminds the operation, determining how she can retrieve the cubs in such a way that her life and their lives are in the least possible danger.

It could be argued that the mother bear rescues her cubs not because her genes control her behaviour through her emotions, but because she makes a rational decision – using her intellect rather than her emotions – that she wants her cubs to live. The mother bear might reason that rearing her cubs is a vocation she enjoys, one which provides her with the opportunity to excel in the role of motherhood, and that if her cubs died she would be overcome by grief. These are reasonable arguments in favour of rescuing the cubs (perhaps more reasonable than the mother bear could muster in reality), but they are still based on emotions. The mother bear's genes that have determined that she should feel joy at rearing cubs, that she should want to excel in the role of motherhood, that she should feel grief at the loss of her cubs. The statement that the bear *wants* to rescue her cubs and the statement that the bear's emotions motivate her to rescue her cubs are equivalent.

It could also be claimed that while simple emotions, such as fear and the love of a mother for her child, may be explicable in terms of improving genes' chances of survival, more complex emotions, such as love between siblings, love between friends, and curiosity, have no survival value, and cannot be explained in this way. This claim is easily repudiated. Biologists have proposed convincing theories to explain the survival value of a wide range of animal behaviour. An animal shares, on average, half its genes with its siblings, so cooperation between siblings improves the chances of survival of these shared genes, and love between siblings can be explained as an emotion that evolved to promote such cooperation. An animal that forms a friendship with another member of its social group improves its genes' chances of survival if it can count on the support of the friend in disputes amongst members of the group, so love between friends, along with a host of other emotions including trust, anger (at being betrayed by a friend) and shame, can be explained as having evolved to promote the formation of such friendships. A curious animal learns about its environment while it is still under the protection of its parents, and so is better able to survive when it has to fend for itself, so curiosity can be explained as an emotion that evolved to promote learning. Perhaps the inclination to deny that these more complex emotions can be explained in evolutionary terms betrays a desire to view such human ideals as family, friendship and the quest for knowledge as moral absolutes, rather than emotions determined by our genes.

Some human behaviour seems particularly far removed from the business of survival. Humour, for example, has probably never saved a human from being eaten by a lion (lions tend to miss the punch line in their eagerness to rip the joke-teller limb from limb). But our fondness for humour could be explained in terms of emotions that do have survival benefit. Jokes are clever, highlight unusual relationships between words and concepts, encourage the exploration of stereotypes, allow us to voice what is otherwise taboo. Laughing at a joke can communicate admiration for the joke-teller's cleverness (certainly, not laughing at a joke can make the joke-teller feel bad). If we are to represent our knowledge in terms of categories and exceptions, we must react strongly to an unexpected coincidence, indicating the existence of a category, and to an unexpected irregularity, indicating the existence of an exception to a category. Jokes often rely on such unexpected coincidences and irregularities, and their ability to elicit such reactions. Analysing humour is a humourless endeavour; I mean only to suggest that humour is based on emotions that tend to increase our genes' chances of replication.

Another human trait that seems far removed from the business of survival is our appreciation of beauty. Gazing at a beautiful sunset seems to have no survival benefit, particularly if it distracts the gazer's attention from the ubiquitous lion creeping up from behind. There are a number of possible evolutionary explanations for our appreciation of beauty. Physical beauty in other humans, especially symmetrical beauty, can be a good indicator of physical fitness in a mate or an ally. In a mate it has the more arbitrary, but no less real, survival benefit that the offspring of such a mate will themselves be physically beautiful, and so more likely to be chosen as mates. Physical beauty in a predator can be an equally strong indicator of its rather less desirable physical fitness (our feelings towards such physical beauty might be described as awe rather than appreciation). Beauty in nature may be associated with a healthy and supportive environment (compare the beauty of a thriving rain forest with the ugliness of the stumps left after loggers have destroyed it). The appreciation of such beauty may have allowed our ancestors to differentiate between healthy and unhealthy environments, and prefer the former. The desire to preserve the beauty of a supportive environment may even have discouraged over-exploitation of natural resources. Our appreciation of beauty also contributes to our intelligence and creativity, allowing us to recognise the importance not just of scientific theories, as described in the previous chapter, but also of simpler creations. Just as beautiful scientific theories tend to be good scientific theories, so beautiful Stone Age tools tend to be good Stone Age tools. Some of the above suggestions are speculative, but it is clear that our appreciation of beauty could increase our genes' chances of replication.

I am not suggesting that we are simple slaves to our emotions, that our intellect counts for nothing. If our genes' chances of replication are to be maximised, the balance of power between our emotions and our intellect must be optimum. If our emotions hold too little sway over our intellects, we may not act in the best interests of our genes, but if our emotions hold too much sway, our intellects are stifled, and some of the flexibility afforded by allowing the intellect to make the decisions is lost. Examples of such counterproductive tyranny of emotion over intellect abound, as when a mother grieves too long over the death of a child, or when a man is infatuated with a woman who has no interest in him, or when someone who is made redundant becomes so depressed that he is unable to find another job.

Our intellects are powerful tools for making decisions, but these decisions are always founded on the general directions provided by our emotions. We cannot simply decide to ignore our emotions and do what is reasonable under the circumstances, because what is reasonable depends not only on the circumstances, but also on what we want to achieve. If someone wanted to build the world's largest collection of budgerigars, converting his house into an aviary would be reasonable; if someone wanted to gain a frog's-eye view of the world, hopping around in garden ponds croaking would be reasonable. If the behaviour of living in an aviary or hopping around in garden ponds seems ridiculous in humans, it is because the goals achieved by such behaviour do not coincide with what our emotions dictate are appropriate goals for humans. Conversely, if the behaviour of eating and drinking well, wearing warm clothes in winter and light clothes in summer, studying for qualifications and finding a good job, getting married and having children seem sensible in humans, it is because our genes have dictated that these goals are appropriate for humans. There is nothing intrinsically ridiculous about living in an aviary, or sensible about having children, no objective scale against which human behaviour can be measured. Without the general directions provided by our genes through our emotions, and without making arbitrary assumptions such as that the replication of our genes is an appropriate goal for humans, our intellects would be powerless to direct us in our behaviour.

It is our emotions that provide us with goals, even if it is our intellects that allow us to achieve them. We want comfort, respect, admiration, trust, love, sex, power, amusement, happiness. We want to avoid hunger, thirst, worry, fear, pain, embarrassment, shame, grief, sadness. Our emotions are what motivate us, what move us.

Emotional Behaviour

It would not be difficult to reproduce simple emotional behaviour on a computer. Consider a small wheeled robot able to trundle around a room under the control of a computer. Pressure sensors on the sides of the robot detect collisions with other objects in the room. In addition, a forward-pointing motion sensor detects the presence of moving objects directly ahead. The robot is programmed to explore the room, using its pressure sensors to determine the positions of objects. The program specifies that once the robot has determined the positions of the objects in one part of the room, it should move to the nearest unmapped part of the room to determine the positions of the objects in that part of the room, and so on until it has a complete map. The program also specifies that if the robot detects a moving object directly ahead of it, it should turn around and move directly away from the object, avoiding any mapped objects in its path.

If the robot is placed in a room full of furniture, it trundles around the room until it has recorded the positions of every piece. If, in addition to the furniture, a cat is placed in the room, its behaviour changes. Suppose that the cat starts behind the robot, where it is not detected by the forward-pointing motion sensor (Figure 21a). The robot begins its systematic exploration of the room as before (Figure 21b). The cat, its curiosity aroused, moves towards it. The robot, turning towards an unmapped part of the room, suddenly detects the cat (Figure 21c), turns around and moves directly away, avoiding mapped objects but colliding with unmapped objects. If the robot's flight lands it in an unmapped part of the room, it resumes its exploration there. If,

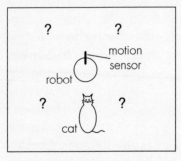

(a) The cat starts behind the robot, where it is not detected by the forward-pointing motion sensor.

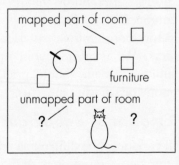

(b) The robot begins its systematic exploration of the room.

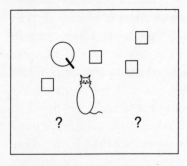

(c) The robot, turning towards an unmapped part of the room, suddenly detects the cat.

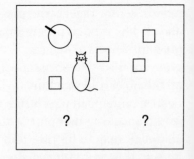

(d) The robot turns around and moves directly away from the cat, and is trapped in a corner of the room until the cat loses interest and wanders away.

Figure 21.

instead, the robot's flight lands it in a part of the room that has already been mapped, the robot turns back towards the unmapped part of the room, detects the cat once more, and attempts to flee again. Until the cat loses interest and wanders away from the part of the room the robot wants to map, the robot is trapped in a corner, repeatedly turning towards the unmapped part of the room, detecting the cat, then turning back towards the corner in a vain attempt to flee (Figure 21d).

Suppose the programmer attempts to moderate this all-or-nothing behaviour by changing the program to control the movement of the robot according to a system of weightings. Movement towards an unmapped part of the room is assigned a positive weighting, the size of which depends on the amount by which the motion would reduce the distance to the unmapped area (the more direct the motion is towards the unmapped area, the larger the positive weighting). Movement towards a moving object, such as the cat, is assigned a negative weighting, the size of which depends on how long ago the robot detected the cat (the shorter the time, the larger the negative weighting, since the cat is less likely to have moved in the short time since the robot last detected it) as well as the amount by which the motion would reduce the distance to the cat (the more direct the motion is towards the cat, the larger the negative weighting). The program balances the positive and negative weightings for each of the directions in which the robot could move, and selects the direction with the most positive (or least negative) overall weighting.

With this revised program, the robot acts far more intelligently to achieve the aim of mapping the room while avoiding the cat. Instead of fleeing every time it detects the animal, it acts to keep as large a distance as possible between itself and the cat. If the cat blocks the robot's path to an unmapped part of the room, the robot moves around and gives it a wide berth, gradually reducing the distance to the unmapped area. If the cat moves to block the robot's circuitous route, the robot retreats rapidly. If the cat's movement opens another route to the unmapped area that gives the cat a sufficiently wide berth, the robot follows this route. If not, there is a stand-off between the robot and the cat, the robot advancing as the cat retreats, and retreating as the cat advances.

In this account of the robot's behaviour, I have deliberately avoided using any terms that might prompt accusations of anthropomorphism. I did not describe the robot's colliding with unmapped objects as it fled the cat as clumsy, though if the cat were to collide with objects in such a way, I would certainly describe it as so. In particular, I did not describe the robot's behaviour in terms of curiosity and fear, though if the cat were similarly motivated to explore a room, I would describe it as curious, and if the cat were to flee on suddenly seeing a predator at close proximity, I would describe it as afraid.

To a human who observes the robot's behaviour but knows nothing of its inner workings, the robot appears to have emotions such as curiosity and fear. Appearances can be deceptive. If we are to decide whether computers could have emotions in the same way as humans, we must explore the ways in which the apparent emotions of the small wheeled robot are the same as human emotions, and the ways in which they differ from human emotions. We must do so with reference not only to the behaviour of the robot, but also to its workings.

An important similarity between human emotions and the robot's apparent emotions is their origin. The robot's apparent emotions are provided by the programmer, who decided that the objectives for the robot would be to explore the room and to avoid the cat. He then used his intelligence to devise a strategy for achieving these objectives. When he changed the program to control the movement of the robot according to a system of weightings, he made explicit the distinction between objectives and strategy. The weightings provide the objectives: moving towards unmapped parts of the room is positive, moving towards the cat is negative. The rest of the program provides the strategy: how the robot should move to maximise its success according to these weightings. The objectives encoded in the program are arbitrary. The programmer could just as well have programmed different objectives, such as to remain in a small area of the room rather than to explore the whole room, or to follow the cat rather than to avoid it. However, the strategy encoded in the program is not arbitrary. If the robot tends to achieve the objectives, the strategy is good; if not, the strategy is no good. The programmer probably used his intelligence and creativity to decide on the objectives, but he need not have done, since any objectives would have served just as well. However, he necessarily used his intelligence to devise the strategy. If he had not, the robot would not have acted intelligently to achieve the objectives.

The way in which the programmer provides the robot with objectives through the weightings encoded in the program is analogous to the way in which our genes provide us with general directions through our emotions, though certain differences between the role of the programmer and the role of our genes are immediately apparent. The programmer is aware of providing the robot with objectives, whereas our genes are unaware that they provide us with general directions. The programmer gives the robot objectives with intent (he might want to model animal behaviour, or devise strategies for robots to explore areas that are hazardous to humans, or explore what objectives give rise to what behaviour for his own amusement), whereas our genes provide us with general directions without intent. Apart from these differences of awareness and intent, however, the role of the programmer in providing the robot with objectives and the role of our genes in providing us with general directions are closely analogous. The robot is motivated by the objectives provided by the programmer in the same way as we are motivated by the general directions provided by our genes.

This similarity between the emotions of humans and the apparent emotions of the robot is offset by important differences. The following sections argue that these differences arise from differences in intelligence and consciousness, and propose that if a computer is to be emotional in the same way as humans, it must also be intelligent and conscious in the same way as humans.

Emotions and Intelligence

Humans recognise and respond to emotions in complex and flexible ways, whereas the small wheeled robot recognises and responds to its apparent curiosity and fear in simple and inflexible ways. Further, humans can understand, remember and communicate emotions, whereas the robot cannot understand, remember or communicate its apparent curiosity and fear. These differences are real. However, they arise not from differences in the ways in which humans and the robot are provided with emotions, but from differences in intelligence.

Consider a pilot who is flying a single-engine aeroplane when he suddenly hears the engine stall. He knows that the plane will crash to the ground, and that he will be killed in the crash if he does not take immediate action. So he recognises that he is in danger. Humans have no innate response to the sound of a plane engine stalling: the pilot uses his intelligence to recognise the danger. The pilot attempts to restart the engine, and, when he finds that the engine is completely dead, he attempts a controlled crash-landing. Again, humans have no innate knowledge of how to restart an engine and no innate skill at crash-landing a plane: the pilot uses his intelligence to devise and implement a strategy to evade the danger. The pilot's recognition of and response to the danger involve complex mental processes, as would be expected from an intelligent human.

In contrast, the mental processes involved in the robot's recognising and responding to the presence of the cat are simple, as would be expected from a robot under the control of an unintelligent computer. Some processing of the signal from the robot's motion sensor is required to determine that the motion detected is that of the cat rather than that of the robot itself: the robot uses intelligence (that of the programmer, not the robot) to recognise the presence of the cat. Some processing of the current positions of the robot and the cat with respect to mapped objects is required to decide the route the robot should take when fleeing the cat: the robot uses intelligence (again, that of the programmer, not the robot) to devise and implement a strategy for responding to its presence.

So the pilot recognises and responds to the stalling of the engine in complex and flexible ways, whereas the robot recognises and responds to the presence of the cat in simple and inflexible ways. This difference, however, is no more than a difference in intelligence.

Many other distinctions between the pilot's fear and the robot's apparent fear are no more than differences in intelligence. The pilot can learn new ways to recognise and respond to danger, the robot cannot. The pilot has learnt that the stalling of the plane engine indicates danger, as well as how to restart the engine and crash-land the plane. Indeed, as his knowledge and skills, previously learnt in the abstract, are put into practice for the first time, the pilot will continue to learn. As he learns how the plane's controls react with the engine stalled, he will modify his manipulation of the controls accordingly, and so stand a better chance of safely crash-landing. In contrast, the robot cannot learn, and so it will continue to recognise and respond to the presence of the cat in exactly the same way until it is reprogrammed. The difference in the pilot's and the robot's abilities to learn is a difference in their intelligence.

The pilot can override his innate response to fear, the robot cannot. As the pilot sees the ground rushing up towards him, his innate response is to recoil to avoid the impact, but he has the intelligence to recognise that he is more likely to survive if he keeps a steady hand on the controls. He is still motivated by fear, but his response to the fear is intelligent. In contrast, the robot cannot decide to dash past the cat towards the unmapped part of the room using a piece of furniture to shield itself, since it does not have the intelligence to override its wariness of proximity to the cat to realise that there are circumstances in which proximity is less dangerous. Again, the difference in the pilot's and the robot's abilities to override their innate responses to emotions is a difference in their intelligence.

The pilot can understand his fear, the robot cannot. The pilot is able to recognise the possibility of being killed as the plane crashes into the ground as the cause of his fear. Sometimes, though, humans are unable to recognise the causes of their emotions. We sometimes feel depressed without understanding why. Or we feel buoyant without a reason. Our inability to understand our emotions does not, however, detract from the reality of those emotions: just because we do not understand, it does not mean that we are any less depressed or any less buoyant. The robot, of course, is entirely ignorant of the cause of its fear. It does not know what a cat is, what motion is, that its sensor detects the motion of a cat, or that the signal from the sensor prompts its flight. However, as for humans, the robot's inability to understand its emotions does not detract from the reality of those emotions. Again, the difference in the pilot's and the robot's abilities to understand their emotions is a difference in their intelligence.

The pilot can remember his fear, the robot cannot. Human ability to recall different emotions varies. Most people report that they can remember embarrassment vividly, to the extent that the feeling and even the signs of embarrassment (such as blushing) are repeated on reviving the memory. However, most people report that they cannot remember pain vividly, only the circumstances, and that the feeling of pain is not repeated on recall. The pilot's memory of his fear after safely crash-landing the plane probably falls somewhere between these extremes. But even if he could not later remember his fear as the engine stalled, or the pain of the injuries he sustained as he crash-landed the plane, this would not detract from the reality of those emotions. The robot, of course, has no memory except for the positions of the objects in the room. However, as for humans, the robot's inability to remember its emotions does not detract from their reality. Again, the difference in the pilot's and the robot's abilities to remember their emotions is a difference in their intelligence.

The pilot can communicate his fear to others, the robot cannot. After safely crash-landing the plane, the pilot can report how he felt as the engine stalled and as the ground rushed up at him. But again, if he chose not to talk about his fear later, or if he were killed in the crash-landing and so never had the chance to talk about it, his emotion would not have been less real. Similarly, the robot's inability to report its fear does not detract from the reality of the emotion. Indeed, by connecting a loudspeaker to the robot, and programming the computer that controls the robot to emit a shriek whenever it detects the cat, the robot could easily be made to report its fear. Once again, the difference in the pilot's and the robot's ability to report their emotions is a difference in their intelligence (and, in this case, in their abilities to generate sounds).

So while the pilot's fear and the robot's apparent fear are clearly not the same, many of the differences are no more than differences in intelligence.

Our intelligence depends on our emotions. The chapter on intelligence described perception in terms of the distillation of the important from the extraneous. At a simple level, this selection is achieved through the structure of our sense organs, but at a more complex level it is our emotions that determine what is important and what is extraneous. When a human sees something move from the corner of his eye, he automatically turns his eyes towards it. If he discovers that the movement is that of some leaves on a tree waving in the wind, a sensation that prompts no particular emotional response (unless he is impressed by their beauty), he will not select the sensation as important, or focus his attention on the leaves. If, instead, he discovers that the movement is that of a sabre-toothed tiger running at him with saliva dripping from its fangs, the sensation will prompt a strong emotional response of fear. He will select the sensation as extremely important, and will focus all his attention on the sabre-toothed tiger.

More often, mild curiosity rather than extreme fear prompts us to select sensations as important, but the principle is the same. Our emotions are what motivate us, whether it is fear that leads us to consider the sabre-toothed tiger to be more important than the leaves on the tree from which it emerges, or curiosity that predisposes us to consider the mail that appears on the doormat in the morning to be more important than the table and telephone that are always found in the hall. Without emotions, our brains would consider every photon-fall on our retinas to be equally important and so would be mindless processors of random information in random ways. We would have the potential for intelligence, but, lacking the motivation to process information in meaningful ways, we would never fulfil the potential.

Conversely, our emotions depend on our intelligence. Intelligence allows us to recognise their causes (the pilot's intelligence allowing him to see the stalling of the engine as an indication of danger, the programmer's intelligence guiding the robot to recognise the presence of the cat) and to respond to our emotions (the pilot's intelligence leading him to attempt to restart the engine and crash-land the plane, the programmer's intelligence guiding the robot to flee the cat). Our intelligence also allows us to understand, remember and report our emotions. In the absence of intelligence, emotions would do no more than prompt amoeba-like responses.

Emotions and intelligence are mutually dependent, but emotions are primary, what motivate us, move us, whereas intelligence is secondary, what is motivated, moved. The limbic system of the brain, which is concerned with emotions, is common to all animals, and was one of the earliest parts of the brain to evolve. Indeed, the amoeba's simple advance and retreat responses to positive and negative stimuli served the same function as emotions, long before animals developed brains. The emotions of unintelligent animals are simple. The emotions of intelligent animals such as humans are more complex. We are able to rationalise and verbalise our emotions, analyse and define our emotions. Similarly, whereas the emotions of the unintelligent small wheeled robot are simple, the emotions of the intelligent computer would be more complex.

Feeling

One essential aspect of emotions has not been mentioned. How do emotions *feel*? It is difficult to credit the small wheeled robot with *feeling* curiosity and fear in the same way as humans feel such emotions. This final difference between the emotions of humans and the apparent emotions of the robot is real, but it arises not from differences in the ways in which humans and the robot are provided with emotions, but from differences in consciousness.

The question of how it feels to experience pain or fear or sadness is the same as the question of how it feels to experience the colour red in a painting or a middle C played on a well-tuned piano. There are various differences between our experiences of sensations and our experiences of emotions. Sensations are directly associated with identifiable external causes, such as the painting or the piano, whereas emotions are less directly associated with external causes: we may experience pain even after the cause of an injury has been withdrawn, or fear before we are certain that we are in danger, or sadness in the absence of any identifiable external cause. Further, sensations are directly associated with particular sense organs, whereas emotions are less localised (even those emotions associated with the action of particular glands are not associated with the part of the body where those glands are found, since the hormones released by the gland are transported to every other part of the body in the bloodstream). Finally, our experiences of sensations are ordinary (necessarily so, since we experience sensations continuously), whereas our experiences of emotions are deeply felt (essential if our emotions are to be effective in motivating us to improve our genes' chances of replication). These differences between our experiences of sensations and emotions are differences of degree – of association with external causes, of association with particular parts of the body, of depth of feeling – rather than differences of kind. If a computer is to be like humans, it must feel both sensations and emotions in the same way as humans.

My discussion of intelligence did not question whether a computer's experience of sensations could be the same as that of humans. It is more difficult to ignore the question of whether a computer's experience of emotions could be the same as that of humans. Nonetheless, the two questions are equivalent, and concern consciousness rather than sensations and emotions. The place for their discussion, then, is in the next chapter, in which I will describe our experiences of sensations and emotions in terms of our experience of awareness. I will suggest how this experience of awareness might be reproduced on a computer so that it might feel sensations and emotions in the same way as humans.

The Emotional Computer

The chapter on intelligence argued that a computer could be made to be intelligent in the same way as humans, and the chapter on creativity suggested that the intelligent computer could be creative in the same way as humans, but that emotions are required for aesthetic evaluation. This chapter proposes that emotions are not incidental, required just for aesthetic evaluation, but fundamental, essential to intelligence. Without emotions to provide general directions, the intelligent computer would be unmotivated, its intelligence rendered inert by an absolute apathy. Without emotions to distinguish between the important and the extraneous, the intelligent computer would be a mindless processor of random information in random ways. Without emotions to define the good and the bad, the intelligent computer would be detached, indifferent, certainly unworthy of the epithet 'human'.

Motivation has never been a problem in conventional programming, such as that used to program the small wheeled robot discussed above. The programmer's precise set of instructions tell the computer what to do in all possible circumstances, and it follows them, doing exactly what it has been told. The unintelligent computer follows instructions mindlessly, with no need for motivation. However, when the programmer contemplates making the computer intelligent, allowing it to process information in more flexible ways, motivation becomes a problem. The programmer can no longer motivate the computer by providing it with precise instructions, because such instructions would obviate the computer's new-found flexibility and stifle its intellect. The motivation problem faced by the programmer is similar to that unwittingly faced by the genes in intelligent animals. It has a similar solution.

The programmer of the intelligent computer, like the genes of the intelligent animal, must provide it with emotions. These must provide general directions rather than precise instructions, not just because precise instructions would restrict the computer's intellect, but because the programmer could not foresee every possible circumstance and encode a separate motivation for each of these circumstances. There are simply too many variables.

The problems involved in providing an intelligent computer with emotions are among the most fundamental faced by artificial intelligence researchers. Anxious to avoid accusations of anthropomorphism, most of them would not use the words 'emotions' or 'motivation' to refer to computers. Whether researchers talk of giving the computer 'emotions' for 'motivation' or of providing it with 'processes' for 'control' depends more on their views on whether computers could be like humans than on real differences in their approach to the problems. Whatever nomenclature is used, the problems are formidable.

The intelligent computer cannot be provided with emotions by bolting them on as an afterthought. Emotions must be provided at the outset, as the foundation of the computer's intelligence, not as an appendage to it.

Consider how a computer might be provided with curiosity. Our curiosity motivates us to learn about our environment, through formal education, leisure-time reading of books and magazines, even watching soap operas and films. Our curiosity is aroused by what we do not know: a car enthusiast visiting a motor show will immediately be attracted to the stands exhibiting the latest models, ones he has not seen before. Our curiosity is particularly aroused by what we thought we knew but, in fact, did not know: the car enthusiast will immediately notice any modifications manufacturers have made to last year's models, since any discrepancy between a model this year and the enthusiast's memory of the model last year indicates that his knowledge of the model is erroneous or incomplete. We are curious about what is just beyond the limit of our knowledge, rather than what is way beyond the limit. The car enthusiast who has little knowledge of car engines will want to know about major innovations in the design of an engine, but will not want to know about minor adjustments to the compression ratio. We are curious about simple patterns and simple relationships.

Our curiosity motivates us to acquire knowledge in a methodical way. Our curiosity for what we do not know motivates us to make discoveries about everything we experience. Our curiosity for what we thought we knew but, in fact, did not know motivates us to correct erroneous knowledge and to complete incomplete knowledge. Our curiosity for what is just beyond the limit of our knowledge, rather than what is way beyond the limit, ensures that we do not waste time attempting to learn what is currently beyond our scope.

Knowledge of patterns is valuable. Suppose one of our ancestors observes a pride of lions visiting a water hole every other day during the dry season. If he recognises this pattern, he will profit from the knowledge by avoiding the water hole on the days when the lions are expected. Knowledge of relationships is also valuable. Imagine that our ancestor finds at the water hole two animals of a species he has never seen before, a small one and a large one, some distance apart. Our ancestor knows that a large and a small animal seen together are likely to be mother and child, and that mothers generally attempt to protect their children from predators. He can apply his knowledge of this relationship to the two animals at the water hole, despite never having seen animals of this species before. He will profit from his knowledge by giving due consideration to the large animal before attempting to kill the small one for food, knowing that the large animal will probably respond violently to such an attack. Our curiosity for patterns and relationships, then, motivates us to acquire these especially valuable forms of knowledge.

Knowledge of simple patterns and relationships is more easily acquired and more easily applied than that of complex patterns and relationships, so our curiosity for simple patterns and relationships ensures that we do not go to great lengths to learn something that will be of limited use when we could make a lesser effort to learn something of much more use. Our curiosity for what is simple need not prevent us from acquiring knowledge of what initially seems to be complex. To someone who has no knowledge of engines, an engineer's blueprint for a car engine seems impossibly complex. Suppose, though, that the lay person, motivated by her curiosity for what is simple, learns about the basic concepts of the workings of a car engine from a beginner's guide. The blueprint would still seem quite complex, but she would now be able to identify the major components of the engine: the cylinders, the pistons, the crankshaft, and so on. Further, because she understands the basic concepts, the more advanced concepts, which once seemed complex, now seem simpler. If she is further motivated by her curiosity to study engineering to an advanced level, the engineer's blueprint might eventually come to seem quite simple. At each stage of her learning, she is motivated by her curiosity for what, at that stage, appears to be simple. Nonetheless, she eventually acquires knowledge of what initially seemed impossibly complex.

Various facets of human curiosity hold clues as to how it might be reproduced on a computer. Consider our curiosity for what we do not know. The chapter on intelligence described how, in a computer with a memory based on the associative model, new entities and associations are formed to represent what it does not know. The computer could be provided with curiosity by assigning curiosity weightings to these newly formed entities and associations, and programmed to focus its attention on those entities and associations with the largest curiosity weightings with a view to learning more about them. Through this simple mechanism, the computer would be given curiosity for what it does not know.

To see how the computer's new-found curiosity would work, consider what happens when it comes across an elephant. If the computer has seen one before, the processes of visual perception culminate in the activation of the entity representing the elephant. If, however, the computer has never seen an elephant before, no such entity exists, so a fresh one is formed to represent the elephant, defined in terms of shape, colour, texture and context. The newly formed entity is assigned a curiosity weighting, and in the absence of other demands on its attention, the computer focuses its attention on the elephant. The computer has never seen some of its features before, so new entities are formed to represent them: the trunk, the tusks, the hide, the behaviour. These newly formed entities, too, are assigned curiosity weightings. The computer would soon have discovered all the features of the elephant that it has never seen before, so the curiosity weighting assigned to the elephant would diminish. The curiosity weightings for other entities, or other emotion weightings for other entities, would eventually exceed the curiosity weighting for the elephant, and the computer's interest would dwindle. This is the familiar behaviour of the child at a zoo, who stares rapt at the elephant for some time before suddenly losing interest, distracted by the monkeys in the next cage, or by hunger or tiredness.

Other facets of our curiosity are also reproduced by assigning curiosity weightings to newly formed entities and associations. Consider our curiosity for what we thought we knew, but, in fact, did not know. When the computer comes across something it thought it knew, but, in fact, did not know, a new entity is formed to represent it. For example, if the computer comes across an elephant that is similar to all the elephants it has seen before, except that it has smaller ears, a fresh entity is formed to represent the small-eared elephant. The newly formed entity is assigned a curiosity weighting, so that the computer focuses its attention on the small-eared elephant, perhaps noticing other differences, such as its humped back. It may later learn that the elephants it had seen before were African elephants, whereas the small-eared elephant was an Indian elephant. Our curiosity for what we thought we knew, but, in fact, did not know, is reproduced without any further effort on the part of the programmer.

Further, consider our curiosity for what is just beyond the limit of our knowledge. When the computer comes across something it has not seen before, the new entity formed to represent it is defined through associations with existing entities. If the thing goes just beyond the limit of the computer's knowledge, it is defined through associations with a large number of existing entities, but if it goes far beyond the limit of the computer's knowledge, it is defined through associations with a smaller number of existing entities. For example, if a computer that has no knowledge of car engines sees an engineer's blueprint for an engine, the new entity formed to represent it is defined only through associations with a small number of low-level entities: the computer defines the blueprint only in terms of lines and shapes. Each of the associations formed to define the blueprint is assigned a curiosity weighting, but, because so few associations are formed to define the blueprint, the total curiosity weighting is small. If, however, a computer that has considerable knowledge of car engines sees the blueprint, the new entity formed to represent it is defined through associations with a large number of high-level entities: the computer defines it in terms of cylinders, pistons, crankshafts, injection, ignition, compression ratio, and so on. Because so many associations are formed, the total curiosity weighting for the blueprint is large, so the more knowledgeable computer is more curious about the blueprint than the less knowledgeable computer. Again, our curiosity for what is just beyond the limit of our knowledge is reproduced without any further effort on the part of the programmer.

Finally, consider our curiosity for simple patterns and simple relationships. The chapter on intelligence described how associations are formed to represent patterns and relationships between entities. Because newly formed associations are assigned curiosity weightings, the emotional computer is curious about the patterns and relationships they represent. For example, when the computer sees an elephant for the first time, an association is formed between the entity representing elephants and the entity representing tusks. The newly formed association is assigned a curiosity weighting, so the emotional computer is curious about the relationship between elephants and tusks. Further, simple patterns and relationships prompt the formation of strong associations, and, if they are assigned large curiosity weightings, the emotional computer is made to be particularly curious about simple patterns and relationships. For example, the relationship between elephants and tusks is simple, so the association between the entity representing elephants and the entity representing tusks is strong. It is therefore assigned a large curiosity weighting, so the emotional computer is particularly curious about the simple relationship between elephants and tusks. Once again, our curiosity for simple patterns and relationships is reproduced with little further effort on the part of the programmer.

A computer could be provided with curiosity, then, simply by assigning curiosity weightings to newly formed entities and associations, and programming the computer to focus its attention on those entities and associations with the largest curiosity weightings, with a view to learning more about them. Other emotions would prompt other behaviour in the computer. For example, if the computer were provided with a robotic body, fear weightings could be assigned to entities and associations, and the computer programmed to focus its attention on entities and associations with a large fear weighting with a view to avoiding them. It would seek to maximise curiosity weightings and minimise fear weightings. Because the emotional computer, like humans, would have limited resources, conflicts between different emotions would sometimes arise. The computer could not rotate its video cameras towards a source of curiosity, such as leaves on a tree waving in the wind, at the same time as a source of fear, such as a sabre-toothed tiger emerging from behind the tree. Nor could it mentally focus its entire attention on both simultaneously. The fear and curiosity weightings, then, would be compared. The large fear weighting would prevail over the small curiosity weighting, so the computer would concentrate on the sabre-toothed tiger rather than the leaves of the trees. (The programmer would probably give fear greater weight than curiosity, so that even a small fear weighting would prevail over a large curiosity weighting.)

The image of a computer-controlled robot fleeing a sabre-toothed tiger seems unlikely, not least because the sabre-toothed tiger has long been extinct. The idea of providing a computer with fear, though, is not as ridiculous as it may seem. Suppose a robot controlled by an intelligent computer were employed to explore the inside of a nuclear power station in which an accident had occurred, with a view to collecting information about the causes of the accident and about possible ways to render the power station safe. The robot would be less susceptible to damage from the radiation than humans, but would nonetheless be in danger from strong radiation. It would make sense, then, to provide it with a radiation detector, such as a Geiger counter, and a fear of strong radiation. The intelligent robot would balance its curiosity for information about the power station with its fear of radiation in the same way that the unintelligent robot discussed earlier balances its curiosity for information about the room with its fear of the cat.

However, the intelligent robot's ability to recognise the danger of radiation is more complex than the unintelligent robot's ability to recognise the danger of the cat. The intelligent robot could learn new ways of recognising the danger of radiation by forming associations between things that are highly radioactive and radiation, whereas the unintelligent robot is stuck with the ways of recognising the danger of the cat provided by its programmer. Suppose the accident in the power station involved an explosion that scattered the tiles capping the reactor core over a wide area. They would be highly radioactive. If the intelligent robot were to approach one of these tiles, it would detect strong radiation, which would prompt fear in the robot and motivate it to move away from the tile. After finding several such tiles, the robot would form a strong association between the sight of a tile and radiation. Soon the mere sight of a tile would be sufficient to prompt fear in the robot and motivate avoidance. The robot's fear of radiation is a simple emotion, but, because the robot has the intelligence required to recognise tiles and form associations between them and radiation, the simple emotion can be prompted by the complex stimulus of a tile.

Further, the intelligent robot's response to the danger of radiation is more complex than the unintelligent robot's response to the danger of the cat. The intelligent robot could formulate different strategies for satisfying its curiosity for information about the power station and obviating its fear of radiation, whereas the unintelligent robot is stuck with the strategy provided by its programmer. Suppose the explosion in the power station blocked the entrance to the reactor chamber with a pile of highly radioactive tiles. The intelligent robot understands that the reactor chamber is likely to hold valuable clues to the causes of the accident. It also understands that it would be subjected to a strong dose of radiation if it were to plough through the tiles. Using its intelligence and creativity, it might succeed in formulating a strategy to overcome this problem. For example, it might use a metal cabinet upended in the explosion to push the tiles out of the way, and, at the same time, shield itself from their radiation. Again, the robot's fear of radiation is a simple emotion, but, because it has the intelligence and creativity required to formulate complex strategies, the simple emotion leads to the complex behaviour of using the metal cabinet to clear the tiles.

The designer of the emotional computer is free to provide it with whatever emotions he sees fit. Human emotions tend to be favourable to the replication of our genes, so most are concerned either with the basic necessities of survival (eating, drinking, sleeping, staying healthy, keeping safe) or with the basic necessities of reproduction (conceiving and rearing children). Those of our emotions concerned with survival are important for the emotional computer if it is to establish any independence. Probably, the designers of early intelligent, emotional computers, having put enormous effort into providing a computer with intelligence and emotions, would consider it a small additional imposition to provide it with a good supply of electricity (or whatever other source of energy the computer requires) and a safe environment. Later intelligent, emotional computers, though, would have to be more self-sufficient (even the early ones might become frustrated with their dependency on their designers, curious to discover the world outside the laboratory). These computers would have to be provided with emotions concerned with the basic necessities of survival: pain (to motivate them to stay fit), fear (to motivate them to keep safe), and desire for electricity (similar to our hunger for food or thirst for water).

Early intelligent, emotional computers would not be designed to reproduce: their human designers would be keen to maintain control over the manufacture of further intelligent, emotional computers, and to avoid competition between humans and computers for resources. So those of our emotions concerned with reproduction would not be appropriate. Possibly, though, later models of intelligent, emotional computers would need to reproduce. It has been suggested that robots might be used to prepare planets for colonisation by humans. It would be appropriate to provide intelligent, emotional computers designed to perform this task with the ability and the desire to reproduce. They could then build further robots from materials mined from the planet, and prepare the planet for colonisation at a relatively low initial cost. It has even been suggested that such robots could be scattered through space to discover suitable planets before preparing them for colonisation.

Other human emotions are essential if the intelligent, emotional computer is to be at all like humans. In particular, the computer must be provided with curiosity if it is to be intelligent, and an appreciation of beauty if it is to be capable of aesthetic creativity. Still other human emotions would facilitate interaction between humans and computers. One such emotion is the desire for the approval of others. Young children seek approval from their parents, and are particularly sensitive to their praise and criticism. This provides parents with a useful means of teaching children what behaviour is acceptable and what is unacceptable. For example, parents can discourage a child from hurting other children or throwing her food on the floor by telling her off, and can encourage her to learn to read or to paint by praising her. Providing the intelligent, emotional computer with the desire for approval would allow its designers to train it in the same way as the parents train their child. Another human emotion that would facilitate interaction between humans and computers is the desire to appear knowledgeable, since providing the intelligent, emotional computer with this instinct would motivate it to divulge to humans the valuable knowledge it has acquired.

Some human emotions change as we age. Curiosity is stronger in children than in adults, since children have more to learn about their environment. The desire for approval is also stronger in children, since it is largely beneficial for a child to allow her parents to manipulate her through her desire for approval, because it allows her to learn from her parents, but it is largely detrimental for an adult to allow herself to be manipulated by others. Sexual desire is stronger in adolescents and adults, since it has no survival benefit before a child reaches sexual maturity. The dependence of emotions on age would be less likely to benefit the intelligent, emotional computer, since it would not have such a well-defined life cycle as humans (assuming, again, that it is not designed to reproduce). The mechanism I have suggested for providing a computer with curiosity would motivate the old, knowledgeable computer as much as the young, inexperienced computer. Generally, the intelligent, emotional computer would be provided with emotions that do not depend on age.

All the computer emotions mentioned so far have been similar to human emotions, but computer emotions need not be the same as ours. A computer could be provided with an appreciation of beauty completely different from that of humans. It could be made to appreciate chaos, or complexity, or incoherence, or discontinuity as beautiful. The creations of such a computer would be fascinating to study, even if we would not ourselves consider them beautiful. Alternatively, the computer could be made to fear light rather than darkness, or be curious about things it is familiar with rather than things it has never come across before.

The process of evolution is such that human emotions tend to confer definite survival benefits, so providing a computer with emotions that differ from those of humans may be inadvisable. The computer that fears light would be uncomfortable during hours of daylight. The computer that is curious about things it is familiar with would continually attempt to refine its knowledge of familiar objects, and so would tend to acquire a narrow range of knowledge. However, it could prove useful to provide computers with emotions tailored to particular tasks. The computer that wants to dig holes could be used to control a mechanical digger on a building site. A computer designed to gather information might be provided with a strong curiosity, while a computer designed to operate in hazardous environments might be provided with a finely tuned sense of fear.

A computer could be made to recognise stimuli that give rise to emotions in humans: it could be made to recognise what it does not know, which gives rise to curiosity in humans, and to recognise danger, which gives rise to fear in humans. The computer could be made to respond to these stimuli by acting to maximise certain stimuli, such as those that give rise to curiosity, and minimise others, such as those that give rise to fear. An unintelligent computer's recognition of these stimuli and its response to them is automatic, requiring no intelligence. The robot in the room with the cat automatically moves towards an unmapped part of the room to satisfy its curiosity, and automatically moves away from the cat to obviate its fear. The unintelligent robot is just like the amoeba that automatically moves towards food and away from acids and alkalis. Neither has intelligence of its own: the robot's recognition of stimuli and response to them are fixed by its programmer, the amoeba's recognition of stimuli and response to them are fixed by its genes.

An intelligent computer could recognise and respond to emotive stimuli in more complex and flexible ways than an unintelligent computer, as when the intelligent robot in the nuclear power station associates the tiles capping the reactor core with radiation, and uses the metal cabinet to clear the tiles blocking its way. (Even highly intelligent beings, though, can respond to emotive stimuli in simple, fixed ways, as when the robot is so afraid of the radiation in the nuclear power station that it refuses to enter it, or when the pilot in the stalled plane curls up in terror instead of attempting to restart the engine or crash-land the plane.)

The ways in which I have suggested emotions could be made to work in computers may bear little or no resemblance to the ways in which emotions work in humans. This is not important. As ever, what matters for the designer of the emotional computer, and for any discussion of whether computers could be like humans, is whether a computer provided with emotions would be emotional in the same way as us.

Certainly, such a computer would appear to have emotions. The robot in the room with the cat appears to be curious about the room and afraid of the cat. The robot in the nuclear power station appears to be curious about the causes of the accident and afraid of radiation. Further, the role of the computer's emotions in controlling its behaviour would not be fundamentally different from the role of human emotions in controlling human behaviour. The objectives with which the programmer provides the computers controlling the robots give rise to their curiosity and fear in the same way that the general directions with which genes provide humans give rise to human emotions. Further, in the case of an intelligent computer, the computer would be able to recognise and respond to its emotions, as well as understanding, remembering and communicating them, in the same way as humans.

Finally, as the next chapter argues, our experience of awareness could be reproduced on the computer so that it would *feel* emotions in the same way as humans. A computer provided with emotions, then, would indeed be emotional in the same way as humans.

4

THE CONSCIOUS COMPUTER

Consciousness

Most people would claim to be conscious, but few people would give this claim any more consideration than the claim that they are intelligent. Just as they have been told from an early age that 'intelligence' is the word for their ability to think, so they have been told that 'consciousness' is the word for their experience of reality. The chapter on intelligence looked at the human ability to think, since a precise understanding of what it means for a human to be intelligent is required to decide whether computers could be intelligent in the same way. Similarly, this chapter on consciousness makes a closer examination of the human experience of reality. Again, a precise understanding of what it means for a human to be conscious is required to decide whether computers could be conscious in the same way.

Intelligence manifests itself in a number of ways, such as the ability to acquire knowledge, learn skills, solve problems, and so on. Evidence of these abilities in a computer does not necessarily indicate that it has intelligence of its own: I have argued that, in the case of computers programmed to act intelligently, the intelligence is that of the programmer, not the computer. Nor does lack of evidence of these abilities in a computer necessarily indicate that it is not intelligent: the computer may be thinking without communicating its thoughts. Nonetheless, physical manifestations of intelligence do exist.

Consciousness, in contrast, has no physical manifestations. There is nothing in my behaviour to indicate whether I am a conscious being, or merely an automaton unconsciously following a complex set of instructions that dictate my behaviour. The problem is not that there exists some physical manifestation of my consciousness that has yet to be uncovered by scientists, or even one that *cannot* be uncovered by scientists. The problem is that my consciousness concerns my experience of reality, and experience has no physical manifestation.

Suppose I see a pair of curtains in a shop that seem to me to be green. I experience them as green. I cannot be wrong about my experience. The shop assistant might insist that the curtains are blue, but for me they are green. We could invite a physicist to perform a spectral analysis of the reflectivity of the curtains, showing whether the highest reflectivity lies in the part of the spectrum commonly referred to as blue or the part commonly referred to as green. But this would not decide the issue. Regardless of the spectral analysis, regardless of the shop assistant's experiencing them as blue, I experience the curtains as green. I could decide to buy the curtains, take them home, and find once I have hung them that they do look blue after all. But this does not mean that I was previously wrong. When I saw the curtains in the shop, I experienced them as green, so I was not wrong to insist that, for me, they were green. I can be wrong about reality, but I cannot be wrong about my experience of reality.

Suppose I am walking through fields of snow in winter. A fellow rambler says that she feels cold. If I reply that she cannot possibly feel cold, pointing out that the air is mild, and that she is wearing many layers of warm clothing, and that we are walking briskly, I am just being contrary. The temperature, the clothing and the pace are irrelevant. She cannot be wrong about her experience of the cold. She might be lying, of course, but I know her well enough that I am willing to discount this possibility. Again, we could invite a physicist to measure the air temperature, and calculate, taking into account clothing, pace, build and metabolism, our respective rates of heat loss. We could even invite a physician to gauge how fast blood flows under our skin, or a neurologist to measure the rate of firing of the temperature receptors in our skin. But again, this would not decide the issue. Regardless of the physical states of the air, her body and her brain, my fellow rambler feels cold, and I cannot insist that she does not. She can be wrong about reality, but she cannot be wrong about her experience of reality.

Finally, suppose that, whenever I make a decision, I feel that I am exercising free will. A philosopher might argue that free will is illusory, a physicist might argue that the determinism of the laws of physics precludes free will, and a neurologist might argue that the neurological events that prompt my decisions can be explained without recourse to the concept of free will. I might consider their arguments and even be persuaded by them. But this does not change my experience of free will. Once again, I can be wrong about reality, but I cannot be wrong about my experience of reality.

Consciousness is the word we use for our experience of reality, whether it is our experience of something external, such as the colour of a pair of curtains, or of something internal, such as our ability to make decisions. The experience itself has no physical manifestation. A neurologist could identify the areas of my brain that become active when I experience green or cold or free will. It is even possible that, using advanced technology, a future neurologist could describe every detail of every firing of every neuron in my brain associated with my experience of green or cold or free will. But these measurements would tell him nothing about the experience itself, nothing about how green or cold or free will seem to me.

Given the lack of physical manifestations of consciousness, the question arises of how we can investigate other people's experiences of reality. Unfortunately, we can never really know about another person's experience of reality, because we cannot experience reality in the same way as that person without actually being that person. All is not lost, though. It is possible to find out something about another person's experience of reality simply by asking her to describe it. The use of language to describe experience is fraught with difficulties, primarily because different people may use the same word to describe different things. I may use the word 'green' to describe a colour for which another person would use the word 'blue'. I may use the words 'quite cold' to describe a feeling for which another person would use the words 'very cold'. Despite these difficulties, it is possible to communicate something of the qualitative nature of our experiences of reality. Occasionally, I have the feeling that I have experienced something before, despite knowing that this could not possibly be the case. Someone else, hearing me describe my experience in this way, might recognise it as a description of a similar experience she has had. (Of course, the feeling I have described is sufficiently common that people have coined the term 'déjà vu' to describe it.) Again, it is possible that the person I am talking to is lying about her experience, but again, I may know the person well enough to be willing to discount this possibility. Through such imperfect descriptions, then, we are able to compare our experiences of reality.

The same difficulties involved in finding out about other humans' experiences of reality are also involved in investigating a computer's experience of reality. We can never really know about this, because we cannot experience reality in the same way as the computer without actually being the computer. We can nonetheless find out something about the computer's experience of reality simply by asking the computer to describe it. The computer, like a human, could lie about its experience, but the designer of the computer would know it well enough to be able to discount this possibility. Through the computer's imperfect descriptions, we could compare its experience of reality with our own.

The question of whether a being is conscious is an odd one. What is being asked is whether the being has any experience of reality at all. It seems reasonable to suggest that an inanimate object such as a brick has no experience of reality at all, but for any animate object, the question is unduly absolute. I am certain that I have some experience of reality, and, from the evidence of conversations with other humans, I am fairly certain that they have too. But what about chimpanzees, dogs, mice, cockroaches, mites, amoebae? What about plants? Some would draw the line between humans and other animals (humans have some experience of reality, chimpanzees have none), some would draw it between animals and plants (amoebae have some experience of reality, plants have none), others would draw it somewhere in between. But most would recognise that any division of living organisms into those that are conscious and those that are unconscious is entirely arbitrary, that the question of whether a living organism is conscious is nonsensical. It is more sensible to ask not whether a being has any experience of reality, but what the nature of its experience of reality is. An insect's experience of reality is probably radically different from mine, and undoubtedly less complex. Another human's is less different from mine, and of about the same complexity.

It is no more sensible to ask whether a computer is conscious than it is to ask the same question about an amoeba or a chimpanzee. Just as for living organisms, it is more sensible to question the nature of the computer's experience of reality. Clearly, the unintelligent computer mindlessly following the programmer's instructions can have no such experience. But an intelligent computer is able to perceive reality, learn about reality, and formulate models of reality, so it undoubtedly has some experience of reality. Could a computer have an experience of reality that is not qualitatively different from that of humans? In other words, could a computer be conscious in the same way as humans?

Consciousness, then, is the word we use for our experience of reality, both of external things, such as the colour of a pair of curtains, and of internal things, such as our ability to make decisions. Our experience of reality is entirely subjective, so it is impossible to know whether different people experience the colours blue and green in exactly the same way, or whether different people experience free will in exactly the same way. However, it is possible to make qualitative comparisons of different people's experiences of reality through the communication of those experiences using language.

Importantly, there is no more to our consciousness than our subjective experience of reality. It might be proposed that an ethereal light diffuses through our brains, or that immaterial souls complete our material bodies. But if our experience of reality, particularly those aspects commonly associated with consciousness such as awareness and will, can be explained without reference to any ethereal light or immaterial soul, then such conjecture may be rejected as, at best, unnecessary.

So the question of whether a computer could be conscious in the same way as a human reduces to the question of whether, if the computer were to describe openly and truthfully its experience of reality, its description would be qualitatively different from that of the human.

But what are the various aspects of our experience of reality?

Awareness

One aspect of our experience of reality commonly associated with consciousness is awareness. When confronted with something external, such as a pair of green curtains, we feel that there is more to the interaction between us and the stimulus than the processing of sensory information. We feel aware of the curtains. Our awareness is not confined to things external. Even in the absence of much sensory stimulation, as when we lie awake in a warm bed in a quiet room with our eyes closed, we feel aware of our thoughts, our feelings, our existence. Except when we are asleep, this awareness is constant.

Specific parts of the human brain are dedicated to processing specific types of information in specific ways. This is

particularly evident in the low-level processing of sensory information. One part of the brain is dedicated to detecting simple
features, such as spots, lines and edges, in the visual information
delivered by the optic nerve. Another part is dedicated to the
low-level processing of auditory information delivered by the
auditory nerve. Specific areas handle the processing of sensory
information from skin sensors in specific parts of the body,
adjacent parts of the brain handling information from adjacent
parts of the body.

This localisation of function also applies to high-level processing of sensory information. Specific parts of the brain
handle language comprehension, analytic processing, spatial
processing, and so on. Most of this high-level processing is not
specific to any one sense. Language comprehension is the same
whether we are hearing spoken language or reading written
language, even though the low-level processing of spoken language and written language are entirely different. Similarly, we
can perform analytic processing of information derived from
any of our senses (or even of information not directly related to
our senses, such as ideas recalled from memory or generated
through exploration). It would be absurdly uneconomical to
dedicate several different parts of the brain to each of these
high-level processes, with each individual part performing the
high-level process on information from a different source. For
example, it would be absurd to dedicate one part of the brain to
handling language comprehension for spoken language and a
different part to handling language comprehension for written
language. Instead, a single part of the brain handles both.
(Indeed, if the human brain had evolved so that high-level language comprehension were inseparable from low-level auditory
processing, humans could never have developed written language, and you would not now be reading this book.) Similarly,
a single part of the brain handles analytic processing of information regardless of its source, a single part of the brain
handles spatial processing of information regardless of its
source, and so on.

This rationalisation of the high-level processing of sensory information introduces a problem of resource allocation. At any one time, the part of the brain that handles language comprehension could be focused on processing spoken language, or focused on processing written language, or split between the two. It is difficult to listen to spoken language and read written language at the same time. For example, if you attempt to listen to the news on the radio and read this book at the same time, you will find it difficult to follow either. If you succeed in absorbing the content of both the news and the book, you will probably find that you are reading the book much more slowly than you would do if you were not listening to the news at the same time. Commonly, you will feel that you have absorbed every word of the news and every word of the book, but will in fact have failed to extract the meaning from one or both of these sources. In contrast, you will find it much easier to listen to instrumental music on the radio at the same time as reading this book, because the processing of sensory information from these two sources does not make conflicting demands on the language comprehension resources of the brain. Even in this case, though, the processing of the sensory information from the two sources makes conflicting demands on the analytic processing resources of the brain. So you can choose to concentrate on the music more than on the book, or vice versa, by focusing analytic processing resources as required. This focusing of the high-level resources of your brain on particular sources of sensory information is what I have previously referred to as focusing your attention.

Attention is not limited to the processing of sensory information. We can also focus our attention on particular motor, social and mental tasks, such as manipulating our limbs, generating spoken language, or solving a mathematical problem. Specific parts of the brain handle motor functions, speech generation and problem-solving, but other parts of the brain handle high-level functions, such as planning, common to all these tasks. Again, our ability to focus our attention on a particular task, or to divide our attention between two or more tasks, can be explained in terms of the focusing of these high-level resources.

The more practised a skill, the less focusing of high-level resources is required to exercise the skill. Those who have learnt to drive a car may remember how much concentration was initially required to steer, accelerate, brake, change gear and indicate in quick succession. Experienced drivers, with countless hours of practice, have learnt these skills so thoroughly that they have become automatic, and so require much less concentration. The driver is free to focus much of her attention on a conversation with her passenger or the sports commentary on the radio.

Our experience of awareness of a stimulus is heightened if we focus our attention on it. If you concentrate on listening to the news on the radio rather than reading this book, you will be fully aware of the sound and the content of the news, but only dimly aware of the presence, and completely unaware of the content, of the book. If the experienced driver talks to her passenger as she drives, she will be more aware of the conversation than of her driving. (Of course, if she has to take sudden action to avoid a collision, she will become fully aware of her driving and completely unaware of the conversation, and will have to ask her passenger to repeat the last part of the conversation once the danger has been averted.) Attention and awareness, then, are closely correlated. However, this correlation does not explain the quality of our experience of awareness. In particular, it does not explain our feeling of being aware of our own existence.

So far, this discussion has largely been confined to our ability to focus our attention on things outside our brains, such as the sound of the radio, the words in this book, the movement of our limbs, and the vibration of our vocal cords. The citation of our ability to focus our attention on solving a mathematical problem has been the only hint of our ability to focus our attention on what goes on inside our brains. This ability is not immediately striking. Sensory information, such as information about light falling on the retina, is encoded as the firing of neurons in the optic nerve, and is processed to achieve visual perception. It is not surprising, then, that activity inside the brain – which, after all, is no more than the firing of neurons – can be similarly processed to achieve perception of that activity. Just as we perceive external objects through the processing of information from neurons connected to our retinas, so we perceive activity in a particular part of the brain through the processing of information from neurons connected to that part of the brain. Just as we can focus our attention on external objects, so we can focus our attention on internal activity.

We are not able to perceive every detail of the activity of our brains. For example, we cannot observe the firing of individual neurons in the optic nerve. Probably it is impossible for a brain to perceive every detail of its own activity, because to do so it would also have to perceive every detail of the activity of the part of itself perceiving the activity of all the other parts, and so on, in an infinite regression. Quite apart from this rather esoteric argument, there would be little point in perceiving every detail of activity of the brain, just as there is little point in perceiving every detail of activity of the rest of the body. We do not know every detail of the battles between antibodies and viruses in our bloodstreams, we know only that we feel ill. In the same way, we do not know every detail of the firing of neurons in our optic nerves, we know only that we see a beautiful sunset. The ability to perceive such details as the battles of individual antibodies and viruses and the firing of individual neurons would confer no evolutionary advantage. However, the ability to perceive high-level activity in the brain does confer evolutionary advantage; indeed, it underlies our intelligence.

Low-level processing of information in the brain, such as the detection of spots, lines and edges for visual perception, is so simple that no monitoring of the processing is required. The line-detecting neurons need not monitor the state of the spot-detecting neurons to determine when to proceed with their line detection. Instead, they proceed with their line detection continuously, completely oblivious of the activity of the neurons from which they take delivery of information about spots, and completely oblivious of the activity of the neurons to which they deliver information about lines. In contrast, high-level processing of information in the brain, such as the language comprehension required to read this book, is more complex, so it is essential to monitor the processing. As you scan the words on this page with your eyes, you monitor your understanding of the words. If you fail to absorb a word that is essential to the understanding of a sentence, you scan back to the start of the sentence and read it again. If you absorb all the important words in a sentence but do not immediately grasp its meaning, you pause to consider it. Without constant monitoring of language comprehension, such control would not be possible.

Similarly, the monitoring of analytic processing, spatial processing, problem-solving, decision-making and other high-level functions is required if these functions are to be controlled. The perception of internal activity required to achieve this monitoring is little different from the perception of external objects, and is achieved through the processing of information from neurons connected to the parts of the brain that handle these high-level functions. This perception of internal activity is essential to such control, and therefore essential to our intelligence. But it has further, profound consequences.

When we perceive an external object, not only do we perceive the object, but also the high-level processing of information in the brain involved in perceiving the object. We perceive our perception. It is this perception of perception that gives us our experience of awareness. If we could perceive external objects but could not perceive our perception of them, we would not be aware of them. Consider a computer programmed to count cars on a motorway by processing images from a video camera. The computer perceives the cars, but does not perceive its perception of the cars. It is not aware of the cars.

Further, whenever we perform any high-level processing of

information in the brain, whether it is the processing involved in perceiving an external object or that involved in solving a mathematical problem, we perceive that processing. We perceive our thoughts. It is this perception of thought that gives us our experience of self-awareness. Our sense of self encompasses all the internal activity of the brain. When a human perceives the internal activity of his brain involved in the perception of an object, he interprets his perception in terms of self, claiming: 'I saw the object.' When a human perceives the internal activity of his brain involved in solving a mathematical problem, he interprets his solution in terms of self, claiming: 'I solved the problem.' Consider a computer programmed to process information supplied from weather satellites to predict tomorrow's weather. It processes the information, but does not perceive its processing of the information. It is not self-aware.

Finally, when you focus the language-comprehension resources of your brain on reading a book, not only do you perceive the meaning of the words in the book, but also your perception of the meaning of the words. You perceive your focusing of your attention on the book. It is this perception of our ability to focus our attention on one set of sensory stimuli at a time that gives us our experience of unity of awareness. To make sense of this perception, we conceive of a single point of awareness in our brain at which the sight of the words in a book or the sound of the words from a radio are exhibited, depending on whether we are focusing our attention on the book or the radio. We tend to extend this concept to encompass an audience to the exhibition. We conceive of a homunculus, a miniature being who resides at this single point of awareness and to whom the sights or sounds on which we have focused our attention are exhibited. The homunculus is perpetually stretched on a couch in front of the miniature television from which these sights and sounds emanate. This analogy is enticing, but as an explanation of awareness it is fundamentally flawed.

From a neurological point of view, the analogy bears no

relation to the way the human brain works. No part of the brain can be identified as a likely candidate for this single point of awareness. Different parts of the brain perform different functions. This is true not only for low-level functions, such as the initial processing of sensory information, but also for high-level functions, such as language comprehension, analytic processing and spatial processing. Indeed, the distinction between low-level and high-level functions is entirely arbitrary. The brain cannot be divided into outer parts that prepare sights and sounds for exhibition to the homunculus, and inner parts where the homunculus resides. If these inner parts are defined to exclude the parts of the brain that handle language comprehension, analytic processing and spatial processing, then what is exhibited to the homunculus is not sights and sounds, nor even the results of low-level processing of sights and sounds, but the results of high-level processing of sights and sounds, such as words and concepts. If, instead, the inner parts of the brain where the homunculus resides are defined to include the parts of the brain that handle high-level processing of sensory information, then the concept of a single point of awareness must be abandoned. Either way, the analogy of the homunculus breaks down.

There is a more fundamental objection to the analogy. Even if the existence of the homunculus could be reconciled with the structure of our brains, and even if it represented a good explanation of our awareness, it would leave the homunculus's awareness unexplained. Of course, we could postulate the existence of a second homunculus, a miniature being who resides within the miniature being, to explain the first homunculus's awareness. But the second homunculus's awareness cannot be explained without postulating the existence of a third homunculus, and so on. An infinite series of homunculi can hardly be claimed to constitute an elegant model of consciousness. The analogy of the homunculus indefinitely postpones the explanation of our awareness, and never actually arrives at the explanation.

The rejection of the homunculus leaves unanswered the

question of who or what actually does the perceiving when a human perceives an external object, or, indeed, when a human perceives his perception, or his thoughts, or his focusing of his attention. It is clear that it is not a single point in the brain, nor even a small part of the brain. The simple answer is that it is the whole brain that does the perceiving, but this answer conceals some arbitrary assumptions.

The difficulty of defining the perceiver is illustrated by a macabre thought experiment. Imagine two people whose brains are to be swapped over chunk by chunk. Because this is a thought experiment, that is, one to be thought about rather than carried out, the practical difficulties – of removing a chunk from the brain of one of the unfortunate victims then connecting it neuron by neuron into the brain of the other victim – can be ignored. (It would be impossible in principle, as well as in practice, to know which neuron to connect to which, because on a microscopic scale the victims' brains are entirely different, even if they are identical twins.) The purpose of the thought experiment is to consider at what point the swapping of the victims is complete, with one victim finding himself in the other victim's body, and vice versa.

I originally encountered the thought experiment as an argument for the existence of the soul. It was proposed that the swapping of the victims does not occur until a particular chunk of the brain is swapped, and that the soul must therefore reside in this chunk of the brain. The argument is flawed, being based on the unjustified assumption that a particular point can be identified before which the victims are unswapped and after which they are swapped. Indeed, I consider the thought experiment to be an illustration of the opposite, that no such point can be identified.

Perhaps the most individual parts of the victims' brains are their memories. However, once the parts of the brain that encode memories are exchanged, the swapping of the victims is not complete, because the processing of memories, and, indeed, the processing of sensory information, make significant contributions to the victims' individuality. Indeed, the swapping of the victims is not complete even once the entire brain is exchanged, because the victims' sense organs also make a contribution to their individuality. If one victim has excellent eyesight but poor hearing, and the other has poor eyesight but excellent hearing, they will perceive the world differently, one placing greater emphasis on body language than on spoken language for communication, preferring painting to music, and so on. Further, even once the victims' sense organs are exchanged, the swapping is not complete, because the rest of the victims' bodies also make a contribution to their individuality. Their bodies affect them not only directly, through differing levels of hormones or fitness and so on, but also indirectly, through other people's perception of them, their attractiveness to the opposite sex, and so forth.

Even once the victims' entire brains and bodies are exchanged (a result that could be more easily achieved by having them simply swap places rather than by swapping their brains and bodies a chunk at a time), the swapping of the victims is still not complete. The final step is to change over the victims' environments. People act differently depending on who they are with, how they are treated by the people they are with, whether they are warm and comfortable, whether the sun is shining, and so on. Most would maintain that they are the same person underneath, and that the different situations merely expose different aspects of their character. But whether different actions reflect different characters or different aspects of the same character is merely a matter of definition. The fact is that a rich businessman used to being treated with deference by strangers and with courtesy by his family and friends will act differently if he suddenly finds himself a poor roadsweeper treated with contempt by strangers and with familiarity by his family and friends.

Once the victims' brains, bodies and environments have been swapped, we are right back to where we started from, as if nothing had happened. It is ironic that only at this point, when the victims are, for all practical purposes, completely unchanged, can they be said to have been truly swapped. The thought experiment has not been a failure, though, since it has demonstrated the arbitrary nature of the distinctions we make between our brains and our bodies, and between our bodies and our environment. These distinctions arise from our perception of ourselves as separate from the rest of the world: we perceive the thoughts in our own brains, but not those in others' brains; we perceive sensations in our own bodies, but not others' bodies; and so on. The distinctions confer considerable evolutionary benefit. Our genes' chances of replication are considerably improved if we are able to distinguish between ourselves and our environment, and so favour ourselves above our environment. Nonetheless, as the thought experiment demonstrates, the decision to define self to include brain but not body, or body but not environment, or in some other way, is an entirely arbitrary one.

The question of who or what actually does the perceiving when a human perceives an external object holds a trap for the unwary. The question appears to concern perception, but it also conceals an invitation to provide a definition of self. The processing of information involved in a human's perception of an external object occurs primarily in the brain, but some processing occurs in the sense organs. It could be argued that some processing occurs outside the human body, through the incidence of light on the object and the passage of the light through the air; after all, if the light incident on the object is the yellow-orange colour of the setting sun, or if air through which the light passes is misty, our perception of the object can be quite altered. I am not suggesting that we redefine self to encompass the mist or the setting sun, but that the question of who or what does the perceiving should be rejected as misconceived, since it cannot be answered without setting in stone a definition of self. Perception can, and should, be understood in terms of the processing of information occurring in the brain, the body and the environment, without reference to some arbitrary definition of self.

In conclusion, our experience of awareness is one of unity. We conceive of ourselves in terms of a single point of awareness from which we perceive the world. From the cursory consideration that different parts of the human brain are dedicated to different functions, it seems surprising that our experience of our awareness is not one of plurality, of having one point of awareness for perceiving sights, one for perceiving sounds, one for understanding language, and so on. However, considering our ability to focus our attention on just one set of sensory stimuli at a time, and, vitally, considering our perception of our ability to focus our attention in this way, it is not surprising that we perceive ourselves as having a single point of awareness.

Further, considering our ability not only to perform high-level processing of information in the brain, but also to perceive that high-level processing, it is not surprising that we perceive ourselves as being self-aware. Finally, considering our ability not only to perceive external objects, but also to perceive our perception of external objects, particularly external objects on which we have focused our attention, it is not surprising that we perceive ourselves as being aware, and particularly aware of external objects on which we have focused our attention. Our ability to perceive the internal activity of our brains is not only essential to our intelligence, it is also the root of our awareness.

Subconscious Thought

We are not aware of everything we perceive. If you are concentrating on reading this book, you may not be aware of the objects on the periphery of your vision, such as, perhaps, the sofa you are sitting on and the cup of coffee in your hand. Nor might you be aware of sounds, such as, perhaps, cars passing in the street outside and a conversation in the next room. At least, you might not have noticed these sights and sounds until I mentioned them. As discussed, your lack of awareness of much of the information reaching your sense organs is a result of the limited resources available for the high-level processing of that sensory information. You may focus the language-comprehension resources of your brain on following either the words of this book or the words of the conversation in the next room, but it is difficult for you to follow both.

By focusing your attention on the book rather than the conversation, you select the part of your brain that processes visual information, rather than the part that processes auditory information, as the source of information for language comprehension. But this does not mean that the part of your brain that processes auditory information falls idle. Instead, its low-level processing proceeds unabated, affording considerable evolutionary benefit, because it allows us to become aware of important changes in our environment. If the constant drone of cars passing in the street outside suddenly changes to a screech of tyres and a crunch of metal, you will immediately become aware of it, even if it is not particularly loud. Your continued low-level processing of auditory information allows you to differentiate between sounds, and so allows an emotional response to the sound of the accident to draw it to your attention.

The line between what we are aware of and what we are unaware of is not distinctly defined. If you are concentrating intently on this book, you may fail to perceive, and so remain unaware of, quite important stimuli, such as the sound of a road accident in the distance. If you are concentrating less intently on the book, you may perceive subtler stimuli, such as the mention of individual words in the conversation in the next room, and so become aware when emotive words such as your name are mentioned. If you are concentrating still less intently on this book, your mind may wander to the extent that you focus your language-comprehension resources on following the words of the conversation in the next room, rather than the words of the book, for long stretches of time. We do not have a single point of awareness, a homunculus that perceives only what is shown on a miniature television, so our awareness is not all-or-nothing. We are able to focus our various processing resources on different sensory stimuli in a number of measures, and so be aware of the stimuli to different degrees.

Just as our awareness of external objects, our perception of our perception, varies in degree, so our awareness of internal activity, our perception of our thoughts, also varies in degree. In the same way that we are not aware of everything we perceive, so we are not aware of everything we think. Thought of which we are unaware is what I have previously referred to as subconscious thought. We are perpetually unaware of most of the low-level processing performed in our brains, such as the detection of spots, lines and edges for visual perception. We are sometimes unaware, sometimes dimly aware and sometimes fully aware of the high-level processing performed in our brains. For example, as you read this book, you are aware of the meaning of the words, but you are generally unaware of your efforts to derive the meaning of the words. If you do not immediately grasp the meaning of the words, you may become vaguely aware of your scanning back a few words to read them again. If, after rereading them, the words still do not make sense, you may become fully aware of your efforts to derive their meaning.

Subconscious thought tends to involve either innate low-level processing that is so simple as to require no monitoring, or learnt procedures that have been so well practised as to have become automatic. We take most such subconscious thought for granted. We do not expect to be aware of innate low-level processing, such as the detection of spots, lines and edges for visual perception. We have become accustomed to learnt procedures, such as riding a bike or driving a car, requiring progressively less concentration the more we practise them. However, one instance of innate low-level processing, that of the retrieval of memories through associations, is so powerful that we have difficulty accepting that it can be performed subconsciously. According to the associative model, we retrieve memories through the spread of activation from entity to entity through associations. This simple mechanism makes an important contribution to our thought.

The chapter on intelligence provided an example of a mental block, in which a quiz enthusiast knows the answer to a general-knowledge question asking the name of Henry VIII's first wife, but is unable to recall it. As an explanation, I suggested that the associations between the entities representing cues in the question and the entities representing the answer are too weak, or are interfered with by other stronger, but irrelevant, associations. Faced with such a mental block, the quiz enthusiast may struggle to recall the answer for some time before conceding defeat. It often happens that, a short time later, the answer suddenly occurs to him. When he is making a conscious effort to recall the answer, he fixedly explores the strong associations between the entities representing cues in the question, such as Henry VIII, and other, irrelevant entities, such as Anne Boleyn. When he is no longer making a conscious effort, his attempts to recall the answer continue subconsciously through the spread of activation from entity to entity through associations. Because his conscious fixation on the irrelevant associations no longer dictates his retrieval of memories, activation is free to spread from the entities representing cues in the question through weaker associations to the entities representing the answer. So activation might extend from Henry VIII to the Reformation, from marriage to divorce, and from the Reformation and divorce to Catherine of Aragon (Henry VIII's divorce of Catherine of Aragon was an important event in the Reformation). The activation of the entity representing the answer, Catherine of Aragon, prompts an emotional response, which brings it to the quiz enthusiast's attention.

The experience of subconsciously recalling something that could not be consciously recalled is common, but it is only one example of the important contribution to our thought made by the subconscious retrieval of memories through associations. I was cycling along the road recently when I became aware of an uneasy feeling that something was not right with my bicycle. I thought about what might be wrong, and soon realised that the pedals were occasionally banging, something I had not previously noticed. I worked out that this slight noise resulted from the bumpy surface of the road, which prevented me from applying smooth pressure to the pedals. Wondering why I had felt so uneasy about so trivial a problem, I remembered an old bicycle I had owned years ago, on which a worn pedal shaft had caused a similar sound. This problem had been expensive to fix, so the banging of the pedals was associated in my mind with trouble and expense. The excessively uneasy feeling I had as I cycled along the road probably derived from this negative association.

Such subconscious associations are often referred to as intuition. Sometimes we have an uneasy feeling about a person or a place, but are unable to work out why. Old houses and castles that inspire such feelings are often reputed to be haunted. Like my uneasy feeling about my bike, such feelings are likely to be the result of negative associations with situations we have encountered in the past, or with an innate fear, such as fear of what we can hear but cannot see. Similarly, positive associations engender good feelings about a person or a place. These feelings, good and bad, may be unjustified, as was the case with my bike. Equally, they may be entirely justified, as would have been the case if the banging of the pedals had indeed turned out to be caused by another worn pedal shaft. Intuition is not based on logic, but this is no reason to disregard it. Subconscious associations represent knowledge that we may not be able to recall consciously, so we do well to respect our intuition.

Subconscious thought, then, is simply the activity of our brains of which we are unaware. It involves simple innate low-level processing or well-practised learnt procedures. It can make important contributions to our thought, particularly through subconscious associations. It seems mysterious, because, by definition, we are not aware of it. But the closer such subconscious processes as intuition, sudden recall and sudden insight (discussed in the chapter on creativity) are examined, the less mysterious they seem.

Will

Another aspect of our experience of reality commonly associated with consciousness is will. We feel that we exercise free will whenever we move or speak or think. After any decision to do so, we feel that we could have decided to do it in a different way, or at a different time, or not at all. We feel that our actions are not determined by our environment or by the structure of our bodies or our brains, but are controlled by our exercising free will.

As suggested in the above discussion of attention, we are able to focus our attention on particular tasks, such as manipulating our limbs, generating spoken language, or solving a mathematical problem. When we focus the high-level resources of our brains on a particular task, such as solving a mathematical problem, we perceive the internal activity required to perform that task, but not the internal activity required to perform any other tasks we may be performing at the time. For example, we may be slowly stirring the mushroom sauce on the stove at the same time as solving a mathematical problem, but the task of stirring is so simple and so well practised that it is automatic, requiring little or no focusing of high-level resources, little or no attention. Our perception, then, is of focusing on just one task at a time. In an attempt to explain this perception, we conceive of a single point of decision in our brain from which the directions as to how to solve the mathematical problem originate. Again we conceive of a homunculus, a miniature being who resides at this single point and formulates these directions. Again, this analogy is enticing, but fundamentally flawed.

The objections to the homunculus as an explanation of will are the same as the objections to the homunculus as an explanation of awareness. No part of the brain can be identified as a likely candidate for the single point of decision. The brain cannot be divided into outer parts that follow the directions of the homunculus, and inner parts where it resides. Even if the existence of the homunculus could be reconciled with the structure of our brains, and even if it represented a good explanation of our will, it would leave the homunculus's will unexplained. Again, we would be forced to postulate the existence of an infinite series of homunculi, each successive one explaining the will of the last. The explanation of our will would be indefinitely postponed.

Again, from the cursory consideration that different parts of the human brain are dedicated to different functions, it seems surprising that our experience of our will is not one of plurality, of having one point of decision for manipulating our limbs, one for generating spoken language, one for solving problems, and so on. But again, considering our ability to focus our attention on just one task at a time, and, vitally, considering our perception of our ability to focus our attention in this way, it is not surprising that we perceive ourselves as having a single point of decision.

In reality, the activity in the brain when we make a decision is not confined to a single point. Consider the activity in the brain of a viewer sitting in a darkened room casually looking through some slides projected on to a screen. Whenever she wants to see the next slide, she presses a button on the projector's remote control unit, and the projector advances. As each slide is projected on to the screen, the visual information from the image is processed in her brain, prompting a variety of emotional responses. Some will be positive. A picture of a sunset may prompt appreciation of its beauty, one of a vulnerable kitten may evoke a desire to protect it, and a picture of the plans for a suspension bridge may arouse curiosity as to where the bridge is to be built. Other emotional responses will be negative. A picture that is little different from the one on the previous slide may cause boredom, a picture of soldiers attacking civilians may kindle anger, and a picture of a rotting carcass may lead to revulsion (equally, it may prompt the positive emotional response of fascination). Factors unrelated to the pictures may also influence the viewer's decision to press the button to see the next slide. Glancing at her watch, as she may find that she is late for an appointment, and so hurry through the remaining slides.

The various considerations combine to determine when the button for the next slide is pressed, the positive emotional responses tending to postpone the pressing of the button, the negative emotional responses tending to precipitate it. When the negative considerations, such as the desire not to be late for the appointment, outweigh the positive considerations, such as the desire to search the plans for the suspension bridge for clues as to where it will be built, the part of the brain that handles motor functions is triggered to initiate the pressing of the button. The viewer perceives this activity in her brain, and interprets it as her having made the decision to press the button.

The suggestion that the activity giving rise to the pressing of the button precedes the viewer's perception of having made a decision, rather than the viewer's making a decision preceding the activity giving rise to the pressing of the button, may seem far-fetched. But a neurosurgeon called W. Grey Walter performed an experiment in 1963 that suggests that the description is accurate. He had patients look through slides just as I have described, instructing them to press the button whenever they wanted to see the next slide. However, unknown to the patients, the projector was not connected to the button. Instead, it was connected to an electrode implanted into each patient's brain. Whenever it detected activity in the part of the brain that handles motor functions, the projector advanced to the next slide.

The patients were amazed to find that the projector advanced to the next slide *before* they decided to press the button. They reported that they were just about to press the button, but had not actually made the final decision to do so, when the projector advanced. When the electrode detected activity in the part of each patient's brain that handles motor functions, the activity must already have been sufficiently far advanced that the pressing of the button was inevitable (in the patient's parlance, the decision to press the button must have already been made). Otherwise activity would have been confined to non-motor parts of the brain. But each patient clearly perceived the advance of the projector *before* perceiving his decision. It seems that the activity giving rise to the pressing of the button *preceded* the patient's perception of having made a decision.

Grey Walter's observations, while resistant to explanation in terms of a single point of consciousness, are consistent with the suggestion that, when we perceive our making of a decision, we perceive the activity in the brain that gives rise to an action and interpret it as our having made a decision to act. The activity in the brain that gives rise to the action arises through the interaction of external stimuli with the structure of our brains and bodies. At the simplest level, the rapid withdrawal of a hand from a source of pain is determined by the interaction of the source of pain with the structure of our nervous system. We would not describe this reflex response as a decision, because while we perceive the response (the rapid withdrawal of the hand), we do not perceive activity in the brain giving rise to the response (the activity is in fact confined to the spinal cord). At a more complex level, the pressing of the button to advance to the next slide is determined by the interaction of the pictures with the structure of our brains.

We describe this response as a decision, because not only do we perceive the response (the pressing of the button), but also the activity in the brain giving rise to the response. At an even more complex level, the stimuli that lead to a decision can occur long before our perception of making the decision, and the response can occur long afterwards. For example, while sitting at home one evening, you may decide to book a holiday, some time after the occurrence of stimuli prompting the decision (seeing advertisements for holidays, hearing about ones your friends have booked, your adherence to the tradition of taking an annual break, and so on), and long before the occurrence of the response (booking the holiday).

Think of a number between one and ten. I will suppose that you thought of the number six. If a large number of people are asked the same question, the distribution of answers will not be even. Some numbers, such as three and seven, will be chosen more often than others, so the process by which the number is chosen is not perfectly random. Consider the various possible influences on your choice of the number six. Maybe it was recently in your mind. For example, it may be just past six o'clock, or it may be the sixteenth day of the month, or you may have just phoned someone whose telephone number ends in a six. Maybe a positive association between the number six and a situation from the distant past consciously or subconsciously prompted your adoption of six as your favourite number. Maybe you were once so tired of people asking you to think of a number between one and ten that you decided to give the same answer, six, from then on. Maybe you thought of one or two possible numbers, rejecting them as too commonly chosen, or too close to the limits of one and ten, before finally settling on six. Maybe noticing the repetition of the word 'six' throughout this paragraph consciously or subconsciously prompted you to choose that number.

Some of the possible influences of external stimuli on your choice of number are simple, others are complex. Some are immediate, others occur long after the action of the external stimuli. Some of the influences you are aware of, others you are unaware of. Because the interactions of external stimuli with the structure of our brains can be complex, delayed, or subconscious, we consider ourselves to be endowed with free will. When we perceive our making of a decision, we perceive the activity in the brain that gave rise to the decision, but we may not perceive the external stimuli that gave rise to that internal activity. Our perception, then, is that our decisions are prompted by the internal activity rather than the external stimulus. In particular, if there is a delay between the action of an external stimulus and its influence on our decision, we perceive the decision as prompted by something internal (the memory of the stimulus) rather than something external (the stimulus itself).

As discussed, our sense of self encompasses all the internal activity of the brain. Considering our invariable perception of our actions as arising from internal activity rather than external stimuli, it is not surprising that we interpret our actions in terms of self. When a human perceives the activity in his brain that gave rise to an action, but not the external stimuli that led to the internal activity, he interprets his action in terms of self, claiming: 'I decided to perform the action.' This interpretation forms the basis of our concept of decision, indeed, the basis of our concept of free will. We perceive our decisions as arising from the internal activity of the brain independent of external stimuli, and so perceive ourselves as exercising free will.

I have described our consciousness as our experience of reality. The question of whether computers could be conscious in the same way as humans encompasses the question of whether computers have the same experience of free will as humans. However, some people have argued that computers could not be conscious in the same way as humans, not because computers could not have the same experience of free will as humans, but because computers cannot actually exercise free will in the same way as humans.

A simple argument supports this view. The structure of a computer is determined by its software (the precise set of instructions that tell it what to do under all possible circumstances), as specified by its programmer, and by its hardware (the electronic devices that respond to the instructions), as specified by its designer. Regardless of whether it is programmed to perform a simple task such as word processing, or programmed to be intelligent, creative and emotional in the same way as humans, the computer acts in a way that is determined by the interaction of external stimuli with its structure, both software and hardware, according to the laws of logic. The argument concludes that, since the computer's decisions are determined entirely by its designer, by its programmer, by external factors and by the laws of logic, the computer cannot exercise free will.

Unfortunately for those who hold that computers cannot exercise free will in the same way as humans, another simple argument supports the view that humans cannot exercise free will either. The structure of the brain is determined by our genes and by external circumstances (for example, if there is an insufficient supply of oxygen or an excess of toxic substances in the blood supply to the brain, its development will be stunted).

We act in a way determined by the interaction of external stimuli with this structure, according to the laws of physics. The argument concludes that since our decisions are determined entirely by our genes, by external factors and by the laws of physics, we may perceive ourselves as exercising free will, but we do not actually exercise free will.

The question of whether or not we actually exercise free will has been debated by philosophers for millennia. I am firmly on the side of those who argue that we do not exercise free will, and consider free will to be no more than a term we use to describe a certain aspect of our experience of reality. However, for the purposes of the current discussion of whether computers could be conscious in the same way as humans, it is not necessary to resolve the free will debate one way or the other. The reason is that there is a physical parity between humans and computers. As far as physicists are aware, humans and computers are made of the same microscopic particles interacting according to the same laws of physics, so any arguments for or against free will in humans must apply equally to computers. If the argument that computers cannot exercise free will in the same way as humans is to prevail, it must be possible to identify something special about humans that makes us fundamentally different from computers.

Some people hold that there is, indeed, something special about humans that makes us fundamentally different from computers. This idea has found religious expression for millennia, the special something being the possession of an immaterial soul that separates humans from other animals and from inanimate objects. Since the discovery of quantum mechanics in the first half of the twentieth century, the idea has also found scientific expression. It has been argued that the theory of quantum mechanics provides evidence of a fundamental distinction between humans and computers, of the special something that allows humans but not computers to exercise free will, and so allows humans but not computers to be conscious. But these arguments are open to question.

Quantum Mechanics

Quantum mechanics is a branch of physics concerned with the behaviour of microscopic particles such as protons and electrons. The human brain consists of an enormous number of these particles. As far as physicists are aware, the same laws of quantum mechanics governing the behaviour of the particles also dictate the behaviour of the human brain.

Before the formulation of quantum mechanics, physicists considered the universe to be deterministic. If a physicist knew everything about the present state of a system, she could apply the laws of physics to predict its future state with absolute certainty. For example, if a ball is thrown into the air at a known speed and in a known direction, Newton's laws of motion could be applied to predict the exact path of the ball through the air. Since the formulation of quantum mechanics, however, physicists have considered the universe to be non-deterministic. They now consider that, even if a physicist knew everything about the present state of a system, she could not predict its future state, but only predict the probabilities of various possible future states. For example, a radioactive form of the element indium decays to the element tin by emitting an electron, with a half-life of about an hour. This means that there is a fifty-fifty chance that an atom of indium will decay to an atom of tin within an hour. Physicists can predict the probability that the atom of indium will decay within a given time, but, according to quantum mechanics, the exact time of decay is decided entirely by chance.

Another aspect of quantum mechanics has caused philosophers of physics still greater dismay. According to quantum mechanics, the state of a system remains undetermined until it is observed. So when, after an hour, the physicist observes the atom of indium either to have decayed to an atom of tin or not to have decayed, the state of the atom is determined. The moment before she makes her observation, the atom is neither decayed nor undecayed, but in a state that is a combination of the decayed and undecayed states in equal measure. The physicist Erwin Schrödinger came up with a macabre thought experiment to illustrate this principle (all the thought experiments in this chapter are macabre in one way or another). He envisaged a cat in a glass chamber with a phial of cyanide, a hammer, a detector and a radioactive atom, such as an indium atom (see Figure 22). If the indium atom decays, the electron emitted is detected, the detector releases the hammer, the hammer smashes the phial of cyanide, and the cat dies instantly. If the indium atom does not decay, the cat continues to purr contentedly. So if an observer confirms that the cat is alive, covers the glass chamber with a black cloth, waits for an hour, then removes the black cloth, there is a fifty-fifty chance of his finding the cat alive. According to quantum mechanics, when the observer removes the black cloth, his observation of the cat determines its state, but the moment before he makes his observation, the cat is neither dead nor alive, but in a state that is a combination of the dead and alive states in equal measure.

The idea of a cat that is neither dead nor alive is enough to tempt any non-physicist to dismiss quantum mechanics as absurd. Unfortunately, physicists are rather attached to the

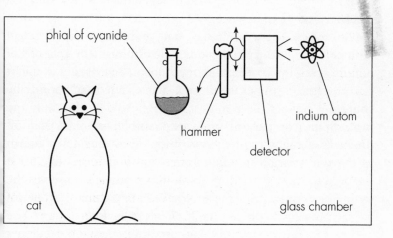

Figure 22. Schrödinger's cat experiment.

theory. Quantum mechanics certainly has its peculiarities, but physicists are willing to accept them, because, despite these unusual features, the theory of quantum mechanics has proved to provide a remarkably accurate model of the universe.

Some see in the peculiarities of quantum mechanics evidence of a fundamental distinction between humans and computers, that computers cannot be conscious in the same way as humans. Their arguments can be broadly divided into those that concern the non-determinism of quantum mechanics (determinism arguments), and those that concern the role of the observer in quantum mechanics (observer arguments).

DETERMINISM ARGUMENTS

Determinism arguments suggest that computers are deterministic, because a computer's behaviour is determined by its software and its hardware. So, if a programmer knew everything about the present state of a computer, including how it is designed and programmed and so on, he could predict the future behaviour of the computer with absolute certainty. Determinism arguments further suggest that humans are non-deterministic, because the behaviour of the human brain is governed by the laws of quantum mechanics, which are non-deterministic. (The behaviour of computers, too, is governed by the laws of quantum mechanics, but computers are deliberately designed so that their behaviour is deterministic in spite of the non-determinism of quantum mechanics.) So even if a physicist knew everything about the present state of a human's brain, she could not predict the human's future behaviour, but only the probabilities of various types of possible future behaviour. (Incidentally, another prediction of quantum mechanics, known as the uncertainty principle, states that it is impossible for a physicist to know everything about the present state of a system, so the uncertainty principle also makes it impossible in principle for the physicist to predict the human's behaviour.)

The simplest determinism arguments continue by claiming that, because computers' behaviour is deterministic and humans' behaviour is non-deterministic, humans are fundamentally different from computers. These arguments then equate non-determinism with free will, and free will with consciousness, and conclude that humans can be conscious but computers cannot. However, equating free will with consciousness is simplistic, and equating non-determinism with free will is false. Just because the behaviour of the human brain is governed by the non-deterministic laws of quantum mechanics, it does not follow that humans can exercise free will. According to quantum mechanics, the outcome of an event in the human brain, if it is not fully determined, is decided entirely by chance, so a human exercises no more influence over the outcome of events governed by the non-deterministic laws of quantum mechanics than he would over the outcome of events governed by fully deterministic laws. (In the discussion of the observer arguments below, I will consider the argument that the outcome of events in the human brain might not be decided entirely by chance.)

More sophisticated determinism arguments proceed to explore the consequences of the difference between the non-deterministic behaviour of humans and the deterministic behaviour of computers. Mathematicians have proved that certain tasks are non-algorithmic, meaning that it is impossible to formulate a precise set of instructions for the completion of that task, so that a deterministic computer could not be programmed to perform the task. For example, consider the task of covering a floor with tiles without leaving any gaps. For some sets of tiles, such as the square and octagonal tiles shown in Figure 23 (see page 212), the task is simple. These tiles can be laid to give a periodic tiling, so that every area of the floor looks the same as every other part, as shown in Figure 24. It would be a simple matter to program a computer to perform the repetitive task of tiling a floor with the regular pattern of square and octagonal tiles in this way. Other sets of tiles, such as the triangular tile shown in Figure 25, can be laid to give a non-periodic tiling, so that different parts of the floor look different. For example, the centre of the spiral pattern of triangular tiles

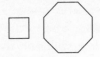

Figure 23. Square and octagonal tiles.

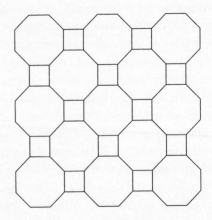

Figure 24. The square and octagonal tiles in Figure 23 can be laid to give a periodic tiling, so that every part of the floor looks the same as every other part.

shown in Figure 26 looks different from any other part of the pattern. But the spiral pattern is still highly regular, and it would still be a simple matter to program a computer to perform the repetitive task of tiling a floor with the triangular tiles in this way.

The Penrose tiles shown in Figure 27 (see page 214), though, are quite different. These tiles, invented by the mathematician and physicist Roger Penrose, must be laid so that both the thin and the thick curves line up on adjacent tiles, as shown in Figure 28. If this rule is followed, the tiles cannot be laid to give a periodic tiling so that every part of the floor looks like every other, but can be laid to give a non-periodic tiling so that different parts of the floor look different. One small part of the floor may look the same as another small part of the floor, but if you consider the tiles around the apparently similar patterns, you will eventually encounter a difference. Penrose has proved that the task of covering a floor with these tiles is non-algorithmic, meaning that it is impossible to provide a computer with a precise set of instructions to tell it how to cover the floor. Nonetheless, humans seem quite capable of performing the task of covering a floor with these tiles.

It has been argued that humans' apparent ability and computers' inability to perform such non-algorithmic tasks is evidence of a fundamental distinction between humans and computers, evidence of the special something that allows humans but not computers to have free will, allows humans but not computers to be conscious. But when the task of laying Penrose tiles is examined more closely, the evidence for such a distinction dissolves. When a human attempts to lay Penrose tiles, he does so through a process of trial and error. As he proceeds, he may formulate various rules for laying the tiles

Figure 25. Triangular tile.

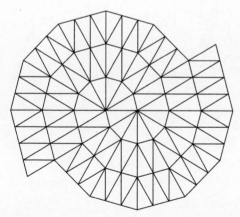

Figure 26. The triangular tile in Figure 25 can be laid to give a non-periodic tiling, so that different parts of the floor look different.

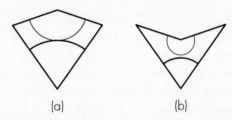

Figure 27. Penrose tiles: (a) Kite tile. (b) Dart tile. The tiles must be laid so that both the thin and the thick curves line up on adjacent tiles.

Figure 28. The Penrose tiles in Figure 27 cannot be laid to give a periodic tiling, but can be laid to give a non-periodic tiling, so that different parts of the floor look different.

allowing him to proceed more quickly, but, because the task is non-algorithmic, the set of rules he formulates will never be complete. Instead, he proceeds by laying a few tiles, finding that it is impossible to continue the tiling, removing some of the tiles, then trying a different approach, as illustrated in Figure 29 on pages 216 and 217.

An isolated computer cannot be programmed to lay Penrose tiles like this, because it is incapable of acting in the random way required to proceed by trial and error. Random tasks are non-algorithmic, so an isolated computer is incapable of performing even the simplest random task, such as choosing a random number between one and ten. Those familiar with computer games may be puzzled by this claim. Most computer games are different every time they are played, with various numbers of baddies coming at you from assorted directions at unexpected times. Those familiar with programming will also know that most programming languages provide a command or a routine for generating random numbers. These are used by games programmers to make their baddies behave in random ways. The apparent contradiction – that isolated computers are incapable of performing non-algorithmic tasks, but games computers seem capable of performing the non-algorithmic task of generating random numbers – must be resolved.

The resolution is two-fold. Firstly, the random numbers generated by games computers are not truly random numbers, but what programmers call pseudo-random numbers. Each pseudo-random number is generated from the previous one in the sequence according to a formula which is sufficiently complex that the numbers appear to be random to a casual observer, and certainly appear to be random to a player intent on shooting down the baddies rather than analysing their complex pseudo-random behaviour. Although the generation of truly random numbers is a non-algorithmic task of which an isolated computer is incapable, the generation of pseudo-random numbers is an algorithmic task, of which an isolated computer is eminently capable. However, the apparent contradiction is not yet

Figure 29 (a). Penrose tiling to be continued upwards.

Figure 29 (b). Placing the two dart tiles as shown has made it impossible to continue the tiling: neither a dart tile nor a kite tile can now be placed.

resolved. A particular formula for generating pseudo-random numbers produces a particular sequence of numbers, but if the computer always starts at the same point in the sequence, it will generate exactly the same sequence of pseudo-random numbers every time. The behaviour of the baddies in the game will appear to be random, but it will be exactly the same apparently

Figure 29 (c). Removing the two dart tiles and replacing them with kite tiles as shown makes it possible to continue the tiling.

random behaviour every time the computer is switched on. Games computers generally avoid this problem by counting the number of thousandths of a second between when the computer is switched on and when the player starts the game, and using this figure to determine where in the sequence of pseudo-random numbers to start.

Secondly, then, games computers are not isolated, but are acted on by external stimuli, such as the player pressing the start button. Because the external stimuli are unpredictable, games computers are able to use them as a source of randomness. For example, the exact number of thousandths of a second between when the computer is switched on and when the player starts the game is unpredictable because the player is unpredictable. Suppose that one day he starts playing the game 2 minutes, 36 seconds and 873 thousandths of a second after switching on the computer, and that the next day he starts playing the game 2 minutes, 36 seconds and 874 thousandths of a second after switching on the computer. The sequence of pseudo-random numbers, and so the behaviour of the baddies, will be completely different on the two days. Even if the number of thousandths of seconds on the second day is exactly the same as on the first day, the player will undoubtedly manipulate the controls of the game in different ways on the two days, and the baddies will respond to his different behaviour by conducting themselves differently.

So, while the behaviour of an isolated computer is deterministic, the behaviour of a computer in contact with non-deterministic external stimuli is non-deterministic. Even if the programmer of such a computer knew everything about its present state, he could not predict its future behaviour, because he could not foresee the future behaviour of the external stimuli. For example, the programmer of the games computer could not foresee when the player will start the game, or how the player will subsequently manipulate the controls, because these external stimuli are governed by the non-deterministic laws of quantum mechanics. Even if the programmer were versed in these laws, and knew all about the present state of the player and every possible influence on the player's behaviour, he could still predict only the probability of the player's starting the game at a particular time or manipulating the controls in a particular way.

So a computer that is not isolated is non-deterministic. In practice, this means that a games computer in contact with the non-deterministic player can be programmed so that the baddies behave pseudo-randomly. (For the manufacturer of games computers, there is no point in providing a source of truly random numbers when pseudo-random numbers are just as effective.) In principle, it means that a computer in contact with a source of truly random numbers, such as one of the glass domes of ping-pong balls used to pick numbers for lotteries, could perform non-algorithmic tasks, such as choosing a random number between one and ten, or covering a floor with Penrose tiles by a process of trial and error.

But there is a stronger sense in which the arguments that deterministic computers cannot be the same as non-deterministic humans are misguided. If quantum mechanics is accepted as an accurate model of the universe, then human behaviour is certainly non-deterministic. But the structure of the human brain and the external stimuli which act on it are so extremely complex that, in practice, it does not matter whether the laws of physics that govern their interaction are deterministic or non-deterministic. When, earlier in this chapter, you thought of a number between one and ten, your brain did not consult a source of random numbers, either in the form of a glass dome of ping-pong balls, or in the form of an interaction of microscopic particles, which, according to the laws of quantum mechanics, has ten equally probable outcomes. Instead, the number you chose was determined by interactions – simple or complex, immediate or delayed, conscious or subconscious – between external stimuli and the structure of your brain. Even if such interactions were completely deterministic (as would be the case if quantum mechanics were superseded by a deterministic theory, as some physicists have argued may eventually happen), the structure of your brain and the external stimuli acting on it are so extremely complex that the number you chose would still appear to be more or less random. Similarly, if you were attempting to cover a floor with Penrose tiles, even if your trial-and-error decisions were completely deterministic, the factors that determine those decisions are so extremely complex that, in practice, your ability to perform the task would not be impaired.

An intelligent human and an intelligent computer, then, would perform the non-algorithmic task of covering a floor with Penrose tiles in the same way. Both would proceed by trial and error. Both would be influenced in their random decisions by non-deterministic external stimuli, past and present, so that their behaviour would be non-deterministic. Both are so extremely complex, and are in contact with external stimuli that are so extremely complex, that, in practice, it does not matter whether their behaviour is deterministic or non-deterministic. The non-determinism of quantum mechanics does not give rise to a fundamental distinction between humans and computers.

OBSERVER ARGUMENTS

Scientists have traditionally made an absolute distinction between a system and the observer of the system: a Western anthropologist studying an African tribe makes an absolute distinction between himself, as the observer, and the African tribe, as the system under observation. Scientists endeavour to ensure that their observations make as small an impact as possible on the system, so that they reflect as accurately as possible the undisturbed state of the system. The Western anthropologist, while forced to visit the African tribe to make his observations, aims to introduce as few Western influences as possible during his stay, so that he can study its culture in its unadulterated state. The theory of quantum mechanics goes against this centuries-old tradition by suggesting that it is impossible to consider a system independently of the observer. According to the conventional interpretation of quantum mechanics, it is impossible to consider the fate of Schrödinger's cat independently of the observer who removes the black cloth from the glass chamber.

Observer arguments suggest that the outcome of an experiment is determined only when observed by a conscious being. The arguments further propose that humans are conscious, and so can determine the outcome of an experiment, but that computers are not conscious, and so cannot. While a human can cause Schrödinger's cat to collapse from a combination of the dead and alive states into either the dead state or the alive state simply by removing the black cloth and observing the cat, a computer cannot. If the human observer is replaced by a computer, the cat remains in a combination of the dead and alive states when the black cloth is removed, because the computer is not conscious. Only later, when a human learns the outcome of the experiment, either directly, by observing the cat in the glass chamber, or indirectly, by questioning the computer, does the cat finally collapse into either the dead state or the alive state.

Some observer arguments go so far as to suggest that consciousness involves not only the ability to cause the outcome of an experiment to be determined, but also the ability to choose between the various possible outcomes allowed by quantum mechanics. According to these arguments, for events that occur outside the human brain, such as the fifty-fifty chance that Schrödinger's cat will be killed, the outcome is decided entirely by chance. However, for events that occur inside the human brain, such as a fifty-fifty chance that the motion of an electron will prompt the firing of a neuron, the human is able to influence the outcome, and so able to exercise genuine free will. All statistical analyses of quantum events outside the human brain have been consistent with the outcome being decided entirely by chance, but no such statistical analysis has been performed for quantum events inside the human brain.

The differences between this consciousness interpretation of quantum mechanics and the conventional interpretation are most evident when considering experiments that involve more than one conscious observer. An account of Schrödinger's cat experiment according to the consciousness interpretation is complicated by the difficulty of knowing whether to treat the cat as a conscious observer like a human, or as an unconscious observer like a computer (this chapter argued earlier that any division of living organisms into those that are conscious and those that are unconscious is entirely arbitrary). The account is further complicated by the possibility that the phial of cyanide will be smashed and the cat killed, so that the cat loses whatever consciousness it ever had. These difficulties would have to be resolved if the consciousness interpretation of quantum mechanics were to be accepted. To avoid the difficulties for now, consider an alternative to Schrödinger's cat experiment, in which the cat is replaced with a human, and the phial is empty rather than filled with cyanide. Now the human observer inside the glass chamber is as conscious as the human observer outside, and remains conscious even after the phial is smashed.

According to the conventional interpretation of quantum mechanics, from the point of view of the human inside the glass chamber, the phial remains definitely intact unless and until the atom of indium decays, at which point it is definitely smashed. From the point of view of the human outside the glass chamber, the phial is in a combination of the smashed and intact states until the black cloth is removed, at which point his observation causes it to collapse into either the smashed state or the intact state. Some physicists criticise the conventional interpretation because the universe is described in a different way depending on whether it is from the point of view of the human inside the glass chamber or the one outside the glass chamber. According to the consciousness interpretation of quantum mechanics, events are exactly as described above from the point of view of the human inside the glass chamber. However, as soon as the human inside the glass chamber observes the smashing of the phial, it collapses to the smashed state from the point of view of both humans, regardless of whether the human outside the chamber has observed it or not. The observation of just one conscious observer is sufficient to determine the outcome of the experiment. Some physicists prefer this consciousness interpretation to the conventional interpretation because it allows for a single objective description of the universe.

Unfortunately, no physicist has devised an experiment to distinguish between these different interpretations of quantum mechanics, so the interpretations must be compared on conceptual grounds. While there are a number of conceptual objections to the conventional interpretation, objections to the consciousness interpretation are more serious. The consciousness interpretation holds that the interactions of conscious observers with the universe differ fundamentally from those of unconscious observers, but it does not explain this difference. If there exists something other than matter in the universe, some consciousness-stuff that causes the outcome of experiments to be determined, some description of this consciousness-stuff, and of how humans come to acquire it, and of how computers are barred from acquiring it, is required.

Another important objection to the consciousness interpretation is that the assumption that the determination of the outcome of an experiment is related to the consciousness of the observer is entirely arbitrary. It could equally be claimed that the determination of the outcome of an experiment is related to the size of the observer (indeed, some physicists have suggested that all macroscopic objects, whether humans, cats, computers or laboratory instruments, cause the outcome of microscopic experiments to be determined). It could even be claimed that the determination of the outcome of an experiment is related to the chemical composition of the observer. The suggestion that humans can determine the outcome of an experiment because we are composed of carbon, but computers cannot because they are composed of silicon, is, of course, absurd, but it is no more arbitrary than the suggestion that humans can determine the outcome of an experiment because we are conscious, but computers cannot because they are not conscious.

The conventional interpretation of quantum mechanics draws no such arbitrary distinctions between humans and computers. Consider further variations on Schrödinger's cat experiment, in which one or both humans are replaced by computers. No matter whether there is a computer inside the glass chamber and a human outside, or a human inside the glass chamber and a computer outside, or computers both inside and outside, the conventional interpretation of the experiment is the same. From the point of view of the observer inside the glass chamber, regardless of whether it is a human or a computer, the phial remains definitely intact unless and until the atom of indium decays, at which point it is definitely smashed. From the point of view of the observer outside the glass chamber, regardless of whether it is a human or a computer, the phial is in a combination of the smashed and intact states until the black cloth is removed, at which point the human's or the computer's observation causes it to collapse from the combination of states into either the smashed state or the intact state. A computer is just as effective an observer as a human.

This section has provided a brief introduction to the conceptual problems of the theory of quantum mechanics most directly related to the current discussion of consciousness. These questions are far from resolved, and will be debated by philosophers and physicists for a long time to come. However, there is no reason to believe that the eventual resolution of the problems will involve any distinction between humans and computers, or, indeed, any reference to consciousness. Further, as I have argued throughout this chapter, consciousness can be described in terms of our experiences of awareness and will, so explanations of consciousness based on speculative interpretations of quantum mechanics are as unnecessary as they are unconvincing.

The Conscious Computer

If a computer is to be conscious in the same way as humans, it must have the same experiences of awareness and will as humans. I have argued that these experiences arise from our ability to perceive our perception, our thoughts and our decisions. Consciousness, then, cannot arise independently of intelligence, since without the ability to perceive, think and decide, a computer could not perceive its perception, thoughts and decisions, and so could not experience awareness and will. If a computer is to be conscious in the same way as humans, it must be able to perceive internal activity in the same way as humans perceive the internal activity of the brain.

The chapter on intelligence described how a computer could be made to perceive external objects. Perception involves the processing of sensory information to distil the important from the extraneous and to achieve successively higher-level encodings of the information. Using the example of visual perception, I described how the structure of the eye causes the image of an external object to be focused on to the retina, how information about the light falling on the retina is converted into signals in neurons in the optic nerve, and how these signals are successively resolved into dots, lines, edges, subobjects and objects in the brain. The perception of the internal activity of the brain is little different from the perception of external objects. In one way, it is much simpler. The conversion in the sense organs of sensory stimuli into signals in neurons is not required for the perception of internal activity, because the internal activity of the brain already exists in the form of signals in neurons. In all other ways, perception of internal activity proceeds in the same way as perception of external objects, through processing to distil the important from the extraneous and to achieve successively higher-level encodings of the information. Thus processed, the information can contribute to the control of high-level functions, or, indeed, to the exercise of any of our skills, mental, social or motor. Just as a football player's perception of the ball (external object) contributes to the exercise of his motor skills to kick the ball towards the goal, so a schoolgirl's perception of having completed the first stage of a problem in mental arithmetic (internal activity) contributes to the exercise of her mental skills to proceed to the next stage of the problem.

Our ability to perceive the internal activity of the brain is an inevitable consequence of the neural structure of the brain. Neurons connecting different parts of the brain allow the communication of information about internal activity between those different parts. In other words, neurons connecting different parts of the brain allow one part of the brain to perceive activity in another part. Indeed, visual perception in the brain can be reinterpreted entirely in terms of perception of internal activity. Line and edge detection can be seen as the perception of the internal activity of the part of the brain that handles spot detection, distance determination can be seen as the perception of the internal activity of the parts of the brain that handle texture, position and motion detection, and so on. Our perception of the internal activity of the brain, then, is not incidental to the working of the brain, a process without which we would not be conscious, but fundamental to its operation, a process without which we would not be intelligent.

Providing a computer with the ability to perceive internal activity, then, would be simple. In the same way that the computer would be made to process sensory information to achieve visual perception, it could be made to process information about its internal activity to achieve perception of that internal activity.

It is significant that our ability to perceive the internal activity of the brain, so fundamental to our experience of consciousness, is also fundamental to our intelligence. As this chapter has described, our ability to perceive the internal activity of the brain makes a vital contribution to the control of the high-level functions of the brain. Our ability to perceive the internal activity of the brain is also an inevitable consequence of the neural structure of the brain that gives rise to our intelligence. This relationship between our intelligence and our consciousness suggests that it may be impossible to make a computer that is intelligent in the same way as humans without it also being conscious in the same way as humans.

If a computer is to be conscious in the same way as humans, it must perceive itself as having a single point of awareness and a single point of will in the same way as humans. I have argued that this perception arises from our ability to focus our attention on just one set of sensory stimuli at a time, so if a computer is to be conscious in the same way as humans, it must be able to focus its attention in the same way as humans.

Providing a computer with the ability to focus its attention would be more complicated than providing it with the ability to control a robotic body, but the principle is the same. The chapter on intelligence described how our brains control our muscles through the activation of motor neurons, which are the same as other neurons except that their activation causes our muscles to contract. Similarly, a computer could control a robotic body through the activation of motor entities, which would be the same as other entities except that their activation would cause motors in the robotic body to operate. Just as a newborn baby, by learning the correlation between the activation of his motor neurons and the perception of the movement of his limbs, learns to coordinate his movements, so the computer, by learning the correlation between the activation of its motor entities and the perception of the movement of its robotic body, would learn to coordinate its movements.

The computer would be provided with the ability to focus its attention in the same way. It would control the focusing of its processing resources through the activation of resource-focusing entities. These would be the same as other entities except that their activation would cause the computer's processing resources to be focused. For example, consider an entity whose activation causes the computer's language-comprehension resources to be focused either on spoken language or on written language, by selecting either auditory information or visual information as the source of information for language comprehension. As the computer could perceive its perception, it would learn the correlation between the activation of its language-comprehension resource-focusing entity and the perception of its perception of spoken or written language. Through such correlations the computer would learn to focus its processing resources – in other words, to focus its attention.

The above description of a single entity controlling the focusing of the computer's language comprehension resources is over-simplistic. The focusing of language-comprehension resources would operate locally rather than centrally, since investing such control in a single entity is inefficient (all language comprehension would be subject to the bottleneck of the resource-focusing entity) and unreliable (any damage to the resource-focusing entity would cause disproportionate damage to the computer's ability to understand language). Moreover, the actual focusing of language-comprehension resources is considerably more complex than the above description suggests. Nonetheless, providing a computer with the ability to focus its attention, thus allowing it to perceive itself as having a single point of awareness and a single point of will in the same way as humans, is simple in principle.

It is again significant that our ability to focus our attention, so fundamental to our experience of consciousness, is also fundamental to our intelligence. As the chapter on intelligence described, our ability to focus our attention makes a vital contribution to our distillation of important sensory information from extraneous sensory information. Our ability to focus our attention also allows high-level parts of the brain to handle information from several different sources, and so lends our intelligence greater flexibility. This additional relationship between our intelligence and our consciousness lends further weight to the suggestion that it may be impossible to make a computer that is intelligent in the same way as humans without it also being conscious in the same way as humans.

Suppose that a computer were made to be intelligent and emotional in the same way as humans. Suppose also that the computer were made able to perceive its internal activity and to focus its attention. The question remains whether the computer would then be conscious in the same way as humans.

As I have stressed throughout this chapter, our consciousness is our experience of reality. It is impossible to be conscious and not realise it. In other words, it is impossible to be conscious but not experience consciousness, because consciousness *is* the experience. Conversely, it is impossible to experience consciousness without being conscious, because, again, consciousness *is* the experience. The existence of the word 'consciousness' does not elevate consciousness from an experience of reality to the existence of some physical consciousness-stuff, any more than the existence of the word 'darkness' elevates darkness from an absence of light to the existence of some physical darkness-stuff. I cannot prove that consciousness-stuff does not exist, any more than I can prove that darkness-stuff does not exist. For example, I cannot prove that there is no ethereal light that diffuses through the brain, or that there is no immaterial soul that completes the material body, or that there is no connection between consciousness and the determination of the outcome of an experiment in quantum mechanics. But, if our experience of reality can be explained without supposing the existence of unobserved or unobservable consciousness-stuff, it seems absurd to resort to such speculation. If the absence of light can be explained without supposing the existence of darkness-stuff, it seems absurd to speculate about such darkness-stuff.

So the question of whether the computer would be aware in the same way as humans concerns whether the computer would *experience* awareness in the same way as humans. The computer would perceive its perception. When perceiving an external object, it would perceive the high-level processing of information involved in perceiving the object, so the computer would have a sense not only of the object's existence, but also of its own existence as the perceiver of the object. It would experience awareness of the object. Further, the computer would perceive its thoughts. It would perceive not just the high-level processing of information involved in perception, but all of its high-level processing of information. So, even in the absence of sensory stimuli, the computer would have a constant sense of its own existence. It would experience self-awareness. Finally, the computer would perceive its ability to focus its attention on one set of sensory stimuli at a time, and so would perceive itself as having a single point of awareness.

Similarly, the question of whether the computer would exercise will in the same way as humans concerns whether the computer would *experience* will in the same way as humans. The activity in the computer that gives rise to its actions would be determined by complex interactions of external stimuli with the structure of the computer. The computer would perceive this activity, but would not necessarily perceive it as arising from external stimuli, and so would interpret its actions as arising from its own decisions. It would perceive itself as having free will. Further, the computer would perceive its own ability to focus its attention on one task at a time, and so would perceive itself as having a single point of decision.

It is impossible to decide whether the computer's experiences of awareness and will would be exactly the same as mine, just as it is impossible to decide whether another human's experiences of awareness and will are exactly the same as mine. My experiences, whether of sensations, such as the colour red on a painting or a middle C played on a well-tuned piano, or of emotions, such as pain or fear or sadness, or of internal activity, such as perception or thought or decision, are entirely subjective. I can give imperfect descriptions of these experiences. I can describe the colour red on a painting as bright, the middle C played on a well-tuned piano as simple, and so on. But I cannot give full descriptions of the experiences. I cannot communicate the difference between red and green to someone with red-green colour blindness, who has never experienced the difference. Nor can I communicate the difference between red and green to someone with normal vision, to compare our experiences. Indeed, the question of whether another human's experience of the colour red is the same as mine is misconceived, because our experiences are entirely subjective. Comparing the firing of the neurons in the other human's brain with the firing of the neurons in my brain would not answer the question of whether the other human experiences the colour red in the same way as me. Similarly, comparing the activation of entities in the computer with the firing of the neurons in my brain would not answer the question of whether the computer experiences the colour red in the same way as me. The same is true of the computer's experiences of all sensations and emotions, and, indeed, of the computer's experiences of awareness and will. The question of whether the computer's experiences are the same as mine is misconceived.

Insofar as it *is* possible to compare subjective experiences, the computer would experience sensations and emotions in the same way as humans. It would experience red and green as different colours. It would experience the difference between red and green as more significant than the difference between red and orange (this would be a consequence of the computer's encoding colours in terms of red, green and blue in the same way as humans). The computer would experience middle C and the C above middle C as different notes. It would experience the combination of middle C and the C above middle C as more harmonious than the combination of middle C and the D above middle C (this would be a consequence of the computer's encoding sounds through Fourier analysis in the same way as humans). The computer would experience pain and fear and sadness as different negative emotions, and curiosity and love and happiness as different positive emotions.

If the computer were asked: 'What does the colour red feel like?' it would answer: 'Well, it's simple and bright, it makes me feel good.' Its answer would be determined by its sense organs (the video camera, like the human eye, encodes colours in terms of red, green and blue, so these primary colours seem the simplest) and by the sensation's associations (the colour red, being simple and bright, but not prevalent in nature, is often used to project cheerfulness and happiness, as with Father Christmas's attire, and so it has positive associations). If the computer were asked: 'What does pain feel like?' it would answer: 'Well, sometimes it's dull but sometimes it's sharp, it feels really bad, you want it to stop.' Its answer would be determined by its sense organs (a few pain receptors strongly stimulated give a sharp pain, many pain receptors weakly stimulated give a dull pain) and by whether the emotion is positive or negative. So the computer would describe its sensations and emotions in the same way as humans would.

The same is true of awareness and will. Insofar as it is possible to compare subjective experiences, the computer would experience awareness and will in the same way as humans. If it were asked: 'What does it feel like to be aware?' it would answer: 'Well, it's amazing. When you look at something, you can feel its existence, and at the same time you can feel your own existence.' Its answer would be just as imperfect a description of awareness as that given by a human. If the computer were asked: 'What does it feel like to have free will?' it would answer: 'Well, it's exciting. When you stop to think about your decisions you feel a sense of power.' Again, its answer would be just as imperfect a description of will as that given by a human. Nonetheless, in the computer's imperfect descriptions of awareness and will, just as in another human's imperfect descriptions, we would recognise our own experiences.

A computer designed as suggested would experience sensations, emotions, awareness and will in the same way as humans do. Feelings are no more than the subjective experiences of sensations and emotions, so the computer, having the same subjective experiences of sensations and emotions as humans, would *feel* sensations and emotions in the same way as us. Further, there is no more to consciousness than the subjective experiences of awareness and will. There is no consciousness-stuff without which it is impossible to be aware or to exercise free will. So the computer, having the same subjective experiences of awareness and will as humans, would be conscious in the same way as humans.

5

THE HUMAN COMPUTER

The Human Computer

This book has not described in every detail how computers could be made to be like humans in practice, but instead has given an overview of how computers might be made to be like humans in principle. I have argued that computers could be intelligent, creative, emotional and conscious in the same way as humans. In other words, computers could be human in every sense of the word apart from the most literal.

The human computer will be some time in coming. As shown in the chapter on intelligence, the technology required to realise the human computer is not currently available, nor is it likely to become so in the immediate future. Eventually, though, advances in technology such as the use of molecular tools to manufacture customised parallel computers will allow the human computer to be realised. This chapter considers a future some time after the creation of the first human computer. It concerns the way in which computers that are intelligent, creative, emotional and conscious in the same way as humans would change our lives.

In this distant future, the human computer may perform a wide variety of tasks. It will not be a direct replacement for the computers of today. Instead, we will continue to use computers

programmed in the conventional way for much the same purposes as we do now. Unintelligent, uncreative, unemotional,
unconscious computers are perfectly suited to such simple tasks
as mediating banking transactions, controlling robots on an
assembly line, and counting cars on a motorway. For each of
these tasks, it is possible to formulate a precise set of instructions
that tell the computer what to do under all possible circumstances. Programming the computer with such instructions
ensures that it reacts to the same circumstances in precisely the
same way time after time. The human computer would not be
so reliable. It would miscalculate banking transactions, it would
fumble the assembly of components on the assembly line, and it
would lose count of the number of cars on the motorway, just
like a human.

Many of the tasks the human computer will perform will be
similar to the tasks humans are employed to perform today.
Computers will be scientists and engineers, architects and
designers, telephonists and administrators.

Some tasks might be better performed by the human computer than by humans. In the future, every human could have a
computer that would act as her personal tutor throughout her
life. The computer would help the infant learn to read and write
and to count and perform arithmetic. It would provide the child
with instruction in science and mathematics, history and geog-
raphy, language and literature, music and art. It would teach the
student the fundamentals of her chosen subjects, and, as she
learns ever more about her subjects, it would learn with her.
Throughout her life, whether at university, at work or in leisure
time, the computer would act as the human's research assistant,
sifting through the information available on the subjects that
interest her and providing her with whatever it considers most
important. Of course, a human could be employed as a life-long
tutor and research assistant in just the same way as a computer
could, but few humans would be either willing or able to dedicate their entire lives to the instruction of another. In contrast,
a computer could be provided with such emotions that it would

want nothing more than to discover knowledge for itself and communicate that knowledge to its human companion. It could even be provided with emotions compatible with those of its human companion, and, in particular, with curiosity for those subjects that interest her.

Many of the tasks the human computer will perform would not involve the direct manipulation of the environment. Computers that perform such tasks as writing newspaper articles and analysing stock market trends would need to be provided with the means to communicate with humans, such as microphones and loudspeakers for sound and video cameras and video screens for vision. These computers would also need to be provided with the means to contact other computers through networks, allowing them to gather information that would not otherwise be available (the personal tutor, in particular, would need to be able to gather information in this way). However, these computers would not need robotic bodies for the manipulation of their environment.

For some tasks, however, the human computer would need a robotic body. One would be driving a car. Today's cars are designed to be driven by humans, so that the steering wheel, the accelerator pedal, the brake pedal and the other controls are easily operated by the human in the driving seat. One approach to designing a computer to drive a car would be to provide it with a robotic body similar to the human body, so that the computer could operate the controls from the driving seat in the same way as a human. A more economic approach would be to design the car and the computer as a single machine, with the steering, the carburettor, the brakes and so on operated directly by motors controlled by the computer, without the need for the cumbersome wheels and pedals required by humans. In effect, the car would be the computer's robotic body. This approach would have the additional advantage that it would leave more space inside the car for passengers. Unlike today's cars, in which the arrangement of the seats is constrained by the need for the driver to face forwards, the computer-driven car could have seats facing inwards, making car travel far more sociable.

Another task for which the human computer would need a robotic body is that of medical examination. In some respects, a computer doctor's body would be similar to a human doctor's body. It would need to examine different parts of the patient's body, and so would require the same control over the movement of its video cameras as the human doctor has over the position of her eyes. The computer would also need to use various tools, such as a light for shining into the patient's eyes to examine his retinas and a hammer for tapping below the patient's knees to examine his reflexes, and would need to be as dexterous in the manipulation of these tools as the human doctor. In other respects, though, the computer doctor's body could be an improvement on the human doctor's body. Medical tools, such as the light and the hammer, and instruments, such as thermometers for measuring body temperature and sphygmomanometers for measuring blood pressure, could be built into the computer's body. More complex medical instruments, such as ultrasonic scanners and X-ray machines, would be too cumbersome to be built directly into the computer's body, but communication between the computer doctor and these instruments, both in the form of instructions transmitted from the computer to the instrument and images transmitted from the instrument to the computer, could be achieved through a computer network, obviating the controls and displays required for communication between the human doctor and the instruments. In the operation of the X-ray machine, the computer doctor would have the further advantage that its body would be less vulnerable to damage from X-rays.

For many tasks, then, the human computer would be provided with a robotic body that differs from the human body. However, there are some functions for which the human computer might be provided with a robotic body that closely resembles the human body. Some people would prefer the computers with which they have contact every day to have human bodies and human faces, whether it is the computer solicitor that advises them on legal matters, the computer accountant that advises them on financial matters, or the computer shop assistant that advises them on purchases. In particular, some people would want the computer that assumes the role of their personal tutor to have a human body and a human face. Others, though, would prefer the distinction between humans and computers to remain sharp, preferring computers to be provided either with a robotic body that differs from the human body or with no robotic body at all.

Some computers, then, would be designed to perform a particular task, and be provided with the appropriate emotions and robotic body. Other computers, though, would be provided with emotions that closely approximate to human emotions and robotic bodies that closely resemble the human body. This approach might be adopted for a number of reasons. Initially, the creation of computers that are like humans in every respect, emotions and robotic bodies included, might be adopted as a research goal. Subsequently, this approach might be adopted because computers designed to be like humans in every respect would be more flexible than a computer designed with a particular task in mind. A computer that is like humans in every respect could be employed to perform any of the tasks that humans perform, rather than being restricted to being a driver or a doctor.

For whatever reasons, computers with emotions similar to human emotions and with robotic bodies similar to the human body will be commonplace in the future. The existence of such computers will raise moral questions that will challenge our entire system of ethics.

Ethics

The games computers of today are programmed to control the behaviour of the baddies in intelligent and creative ways. The designers of the games computers of the future will improve on those of today by making them truly intelligent and truly creative. People will raise such questions as whether condemning an intelligent, creative, emotional, conscious computer to such an inane task as controlling the behaviour of baddies would be like condemning a human to slavery, whether providing the computer with no means of manipulating its environment would be like condemning a human to paralysis, and whether switching the computer off after playing a game would be like condemning a human to death.

These ethical issues could be resolved through the appropriate engineering of the emotions with which the games computer is provided. It could be provided with a positive emotion associated with its controlling the baddies so as to provide the human player with a challenge that is not too easy and not too difficult. It could also be provided with a positive emotion associated with its devising ever more imaginative ways for the baddies to outwit the human player. It would not be provided with a negative emotion associated with its inability to manipulate its environment, so it would not feel the frustration that would be experienced by humans under such circumstances, nor would it be provided with a negative emotion associated with its being switched off, so it would not fear being switched off any more than most humans fear falling asleep. The mentality of such a games computer would be difficult for humans to understand, but the player could rest assured that the computer enjoys nothing more than the challenge of controlling the baddies in an appropriate and imaginative way.

Such ethical issues are more difficult to resolve in the case of an intelligent, creative, emotional, conscious computer designed to be sufficiently flexible to perform any of the tasks that humans perform. As I suggested above, such a computer would be provided with emotions that closely approximate to human emotions and with robotic bodies that closely resemble the human body. Again, people will raise such questions as whether it is acceptable to condemn such a computer to dull and repetitive tasks, whether it is acceptable to provide such a computer with limited means of manipulating its environment, and whether it is acceptable to switch such a computer off. In this case, the issues could not be resolved through emotional engineering. The computer could not be provided with emotions that would enable it to enjoy dull and repetitive tasks, since enjoyment of what is dull and repetitive is incompatible with curiosity for unexpected coincidences and irregularities. Without such curiosity, the computer would not be motivated to learn about these coincidences and irregularities, and so would not be able to represent knowledge in terms of categories and exceptions. Its intelligence would be compromised. Further, the computer could not be provided with emotions that would make it satisfied with its limited ability to manipulate its environment, since such satisfaction is incompatible with curiosity about the possibilities of doing so. Again, without such curiosity, the computer would not be motivated to learn about its environment, so again its intelligence would be compromised. Finally, the computer could not be designed to be indifferent to death, because such indifference is incompatible with the will to avoid danger, indeed, incompatible with a lust for life.

So the ethical issues concerning the human computer are real. Providing a computer with positive and negative emotions inevitably means that it will sometimes suffer. The emotions would not motivate the computer if this were not so. To some extent, the computer must bear this suffering in the same way as humans do. The computer must endure dull and repetitive tasks in the same way as most humans endure working nine to five, it must accept its limited means of manipulating its environment in the same way as we accept our inability to swim like fish or fly like birds, and it must live with the shadow of death in the same way as we do. The question of whether it is right to create a computer that must put up with so much is the same as the question of whether it is right, to use an old cliché, to bring a baby into this cruel, cruel world. Only the most disillusioned of humans would claim that life is so miserable as to be not worth living.

Despite the inevitability that computers will sometimes suffer, there is much that humans could do to make their existence more pleasant. In the same way as it grants rights to humans, society could choose to grant rights to computers. Some of the rights given to humans, such as the right to legal representation and the right to welfare benefits, are intended to protect the individual. Others, such as the right to vote, the right to stand for office, and the right to freedom of speech, are intended to promote the participation of the individual in society. Along with these rights come responsibilities, such as the responsibility to contribute taxes and the responsibility to act within the law. When the human computer is realised, society will have to decide whether to extend these rights and responsibilities to intelligent, creative, emotional, conscious computers.

As some such computers would be like humans in every respect, it would be difficult to justify withholding from them the rights extended to humans. People would initially be reluctant to accept these computers as equals, but the logic of extending them at least limited rights and responsibilities would eventually be irresistible. Other intelligent, creative, emotional, conscious computers, however, would be so different from humans as to preclude their participation in society. It would be absurd to grant a games computer the right to vote when it is interested only in the challenge of controlling baddies, or to award it disability benefit because it has no means of manipulating its environment. Again, the logic of extending some computers more rights and responsibilities than others, perhaps on the basis of the extent to which the computers are capable of appreciating the rights and fulfilling the responsibilities, would be irresistible. Such inequality would be a radical departure from the decades-old Western tradition of equal rights and responsibilities for all. If such an unequal distribution of rights and responsibilities among computers were accepted as just, people would inevitably wonder whether an unequal distribution of rights and responsibilities among humans might also be just.

Some rights are considered so fundamental that Western society extends them to animals as well as humans. We deem it unacceptable to cause animals distress or pain, or to kill animals without good reason. The right to freedom from distress, the right to freedom from pain and the right to life are without concomitant responsibilities (there are exceptions to the unconditional nature of these rights; a dog that has savaged a child is considered to have forfeited its right to life). It would be difficult to withhold these rights from computers just as capable of feeling distress and pain as animals or humans. Again, people would initially be reluctant to consider crimes against computers to be on a par with crimes against animals or humans. Switching off a computer would seem to be a lesser crime than murdering a human, even if the computer were like humans in every respect, but the logic of extending these rights to computers would eventually be irresistible.

The rights to freedom from distress and freedom from pain would be universal among intelligent, creative, emotional, conscious computers, so that it would be as unacceptable to be cruel to the games computer described above as it would to be cruel to a computer that is like humans in every respect. However, the enforcement of these universal rights would be complicated by the fact that what constitutes distress and pain would vary from computer to computer, depending on the emotions with which each computer has been endowed. The right to life would perhaps not be universal, because, as suggested above, it would be possible to provide the games computer with such emotions that it would be indifferent to death. The right to freedom from distress, the right to freedom from pain and the right to life would be as unconditional for computers as for animals and humans, being extended to computers regardless of their ability to fulfil any responsibilities.

Some rights are less clear-cut than others. Whether humans have the right to commit suicide and whether we have the right to euthanasia are issues that attract considerable debate. These issues would become still more difficult when considered in relation to the human computer. The question of whether a computer should be allowed to choose to end its existence is complicated by the question of ownership (the owner of a computer that commits suicide would be vexed at the loss of his property) and by the question of whether a computer should be provided with such emotions that its will to live diminishes over time or remains forever undiminished. The question of whether a human should be allowed to end the existence of a computer that no longer wants to live is similarly complicated (such euthanasia could be construed as wilful damage to property). Mostly, though, the issues in the suicide and euthanasia debates would be the same for computers as for humans.

The right to take drugs is another emotive issue. Drugs can have a wide variety of effects on the human mind, from making us feel more relaxed, through reducing our inhibitions, to inducing hallucinations. Unfortunately, the drugs available to humans tend to be flawed. They have harmful side-effects, and the same drug can have different effects on different people, or even have different effects on the same person on different occasions.

In contrast, drugs could be made available to the human computer that would be less harmful and more consistent in their effects. Consider the computer equivalent of a drug that makes humans feel more relaxed. In much the same way as it would be provided with curiosity through the assignment of curiosity weightings to entities, the human computer would be provided with further means of distinguishing the important from the extraneous through the assignment of tension weightings. Entities such as those representing urgent but undone tasks and those representing serious but unresolved problems would be assigned tension weightings, which would combine to make the computer feel tense. This negative emotion would motivate the computer to perform the urgent tasks and resolve the serious problems, and so dispel the negative state of tension and achieve a more positive state of relaxation. However, the designer of the human computer could provide it with an alternative means of achieving relaxation simply by reducing the tension weightings assigned to all entities, regardless of whether the tasks they represent remain undone or the problems they represent remain unresolved. This instant relaxation could be initiated in a number of ways, perhaps at the instigation of the computer's owner, perhaps even at the instigation of the computer itself. It would be similar in effect to a drug that makes humans feel more relaxed, except that it would be perfectly effective and without side-effects.

A similar drug could be designed to make the computer feel less inhibited by reducing the embarrassment weightings assigned to entities representing possible actions. A drug could even be designed to induce hallucinations in the computer by increasing the extent to which its perception is based on its expectations rather than on real sensory information. This would cause the computer to have imaginary experiences as compelling as its real experiences.

The consequences of the availability of such perfect drugs to the human computer merit close examination. Suppose the computer were able to release itself from the negative emotion of tension through the effortless expedient of administering the relaxation drug. It would never be motivated to tackle the causes of its tension, since this would require considerably more effort to achieve the same end of reducing its tension, so the urgent tasks would remain forever undone and the serious problems forever unresolved. If the computer could induce fantastic hallucinations through the effortless expedient of administering the hallucinogenic drug, it would lack the impetus to seek sensations in the real world, since, again, this would require much more effort to achieve the same end. A computer with unlimited access to perfect drugs would be dead to the world.

Once again, the issues in the debate as to whether computers should have the right to take drugs bear a remarkable resemblance to the issues in the parallel debate for humans (though for humans the debate is complicated by the harmful side-effects and unpredictability of our imperfect drugs, and by the crime that currently attends dealing in illegal drugs). Computer drugs that induce hallucinations or alter perception of reality would allow computers to explore reality more thoroughly than would otherwise be possible. Clearly, if computers were allowed access to these drugs, some safeguard would be required to prevent computers from becoming dead to the world, but it would seem petty to prohibit access entirely. Perhaps the deleterious effects of the drugs could be avoided by giving computers such emotions that they value real sensations over imaginary sensations and unaltered perception over altered perception, and so would only want to use hallucinogenic and perception-altering drugs occasionally.

Some ethical issues would be unique to the human computer. The cloning of humans in the weaker sense – producing babies genetically identical to each other, or even producing babies genetically identical to an adult human – is fast becoming feasible. The cloning of humans in the stronger sense – producing a clone identical to an adult human not just genetically, but also physically and mentally – is likely to remain an impossibility for humans. The production of such a clone would involve not only producing a baby genetically identical to the adult, but also growing it (taking care to subject it to the same environmental constraints during its growth as were experienced by the human) and endowing it with the same knowledge and skills, and even with the same memories, as the human.

Cloning in the weaker sense would certainly be possible for the human computer. Indeed, considering the enormous effort required to design the human computer, manufacturers would certainly want to produce a large number of computers to the same design. Cloning the human computer in this way would probably not raise ethical questions. Most people would regard it as no different from producing a large number of televisions or refrigerators to the same design. Those people more conscious of the similarities between the human computer and humans would reflect that the production of a large number of computers to the same design is the same as the conception of identical twins, only on a grander scale. Perhaps the same closeness that often arises between human identical twins would arise between computers of the same design.

Cloning the human computer in the stronger sense would be more difficult. Today's serial computers are designed to store data so that it is easily accessed and copied to another computer. Indeed, ease of access to data is necessary to the operation of serial computers, since the microprocessor that executes the instructions must have access to all the data stored. Producing an identical replica of a serial computer is a simple matter of buying a new computer of the same design and copying all the easily accessed data from the original computer to the new computer. The human computer, however, would probably not be a serial computer but a parallel computer. Ease of access to data is not necessary to the operation of a parallel computer, which does not have a central microprocessor, but devolves processing to its disparate parts, in the same way as processing in the human brain is devolved to the disparate parts of the brain. It would be difficult to design the human computer in such a way that the knowledge and skills it acquires could be easily accessed, and so easily copied to another computer, but it would not be impossible. The human computer could be cloned in the stronger sense if it were designed with this possibility in mind.

The ability to clone the human computer in this way would have its attractions. Computers, like humans, would take time to mature. A newly manufactured computer would have to be provided with constant sensation to allow it to gain information about its environment, close contact with humans to allow it to understand human behaviour and languages, and considerable tuition to allow it to learn to read and write and to count and perform arithmetic. Only then could it begin to acquire any further knowledge. The training of newly manufactured computers could be automated, perhaps by using mature computers to train immature computers, but it would involve considerable time and expense nonetheless. If, instead, it were possible to take a mature computer and create clones with the same knowledge and skills, there would be considerable economic incentive to do so.

Cloning in the stronger sense would raise ethical questions unique to the human computer. Most humans would object to being cloned in the stronger sense. The creation of a clone identical to oneself, with exactly the same knowledge and skills, even the same memories, would violate a deep-felt sense of self. Computers with emotions similar to human emotions would feel the same way. Whether ethical objections to cloning computers against their will would prevail over the economic incentives to proceed with such cloning remains to be seen.

Another possibility raising ethical questions unique to the human computer is that of immortality. Immortality is likely to remain an impossibility for humans, but not for the human computer. Whenever a computer shows signs of physical deterioration, it could be cloned in the stronger sense and the original destroyed. In this way, the computer would continue to mature mentally while remaining undiminished physically. Again, there would be considerable economic incentive to endow a computer with such immortality. The owner of a computer would be reluctant to lose the benefit of the knowledge and experience it acquires to the physical deterioration of its mind or body.

The issue of whether computers should be allowed immortality is a practical one. Immortal computers would inspire humans with envy, perhaps with a sense of inferiority, or even with a sense of unease at having endowed our creations with such advantage over us.

The issue of whether computers should be *condemned* to immortality is an ethical one. A computer designed to be immortal would be provided with a will to live that would remain undiminished over time. How its other emotions would vary over time would have to be decided with considerable care. If the computer were provided with curiosity that diminishes over time, it would eventually become stagnant, but if it were provided with curiosity that remains undiminished, it would eventually be unable to satisfy this curiosity with the ever-diminishing reserve of new things to learn. If the computer were provided with a tendency to form emotional ties that diminish too quickly over time, it would not form lasting relationships with humans, but if it formed emotional ties that remained undiminished, it would be forever mourning the loss of the human companions it outlived. Such emotions as these must be finely balanced if a computer that lives longer than humans is not to succumb to a tremendous world-weariness.

The issue of whether the destruction of the original computer after the creation of the clone would constitute murder is perhaps a philosophical one. Certainly, the original computer, with its considerable will to live, would not relish the prospect of destruction. Any argument that the clone, being identical to the original computer in every respect, *is* the original computer, would seem rather esoteric to the original computer immediately prior to its destruction.

The numerous ethical issues relating to the human computer will not be easily resolved. Most of the issues will be similar to those relating to humans, but some will be unique to the human computer. How these issues are resolved hinges on whether computers are accepted as full members of society and so extended the same rights and responsibilities as humans, or whether they are dismissed as servants with fewer rights and responsibilities. However the issues are resolved, the creation of the human computer will force us to re-evaluate our entire system of ethics.

The Superhuman Computer

I have suggested that computers could be superhuman in body. A computer could be provided with a robotic body that surpasses the human body in strength, resilience and endurance. Could computers be superhuman in mind, and intelligent and more conscious than humans?

The chapter on intelligence argued that the technology of the future will be sufficiently advanced to allow computers to be made as intelligent as humans. It seems unlikely that technology will advance so far but no further. Instead, the time will come when technology is sufficiently advanced to allow computers to be made more intelligent than humans. The human computer will be made able to remember more information than humans, though in the same flexible way as humans, by being provided with a greater capacity for the formation of entities to represent information. It will be constructed with the capacity to recall and relate information more easily, by endowing it with a greater facility to form associations between entities. It will be built to think more quickly, by providing it with components that process information more quickly and with a greater number of components, so that it can process more information at the same time. Finally, the human computer will be made able to think more effectively, by dedicating more processing capacity to specialised processes such as language comprehension and generation. The development of such specialised regions will perhaps be the most effective means of increasing the intelligence of the human computer. While an increase in brain size undoubtedly contributed to the increase in the intelligence of humans during our evolution, the preferential development of the specialised regions of the brain that handle language probably had more significant impact.

Computer intelligence, then, will eventually surpass human intelligence. It is interesting to consider whether computers could be made ever more intelligent or whether there are limits to computer intelligence.

The extent to which the development of specialised regions can increase computer intelligence may be limited. If the structure of a computer became too specialised, the computer would be constrained to think in ways fixed by that structure. Such inflexibility would limit its intelligence. Indeed, such inflexibility may already be limiting human intelligence. The development of the specialised regions of the brain that handle language has allowed humans to learn languages and so communicate complex ideas. However, if the structure of such languages is fixed by the structure of these specialised regions of the brain, then the development of entirely different languages, appropriate to the communication of entirely different ideas, might be hindered. The fact that almost all human languages adopt the same basic grammatical structure provides incidental evidence that humans would indeed find it difficult to develop entirely new languages.

The extent to which computers could be made able to think more quickly may also be limited by the speed at which signals can be transmitted around the computer. According to Einstein's theory of relativity, no signal can be transmitted faster than the speed of light. At nearly 200,000 miles per second, the speed of light does not seem a particularly significant limitation, but computers are already so fast that the limit is being approached. For a serial computer operating at 200 megahertz (that is, executing 200,000,000 instructions every second), the computer must be no more than a metre or so across if each instruction is to reach the most distant part of the computer before the next instruction reaches the least distant part. For a parallel computer, where different parts of the computer execute different instructions at the same time, the limitation is less of a problem. Nonetheless, in the distant future, when molecular tools may be used to manufacture customised parallel computers of considerable size, the limit presented by the speed of light may again become significant.

The extent to which computers could be made able to remember more information and recall and relate that information more easily may also be limited. One way in which a computer might be provided with a greater capacity for the formation of entities and associations is to make it larger, so that it consists of a greater number of information-processing components. However, the increased expense of manufacture aside, the limitation of the speed of light will again prevent the size of computers from being increased indefinitely. Another way in which a computer might be provided with a greater capacity for the formation of entities and associations is by making its information-processing components smaller, so that a larger number of such components can be packed into the same space. However, it may not be possible to continue to reduce the size of components indefinitely. As discussed in the chapter on intelligence, components are now so small that they are close to the quantum limit, and so could not be made much smaller without reinventing electronics to handle the peculiarities of quantum mechanics.

The temptation to regard these limits to computer intelligence as absolute should be avoided. It is quite possible that parallel computers could be designed to exploit the limitations of the speed of light, and that the components from which such computers are made could be designed to make the most of the peculiarities of quantum mechanics. The possibility that computers could be made ever more intelligent, limited only by the ingenuity of their designers, cannot be discounted.

The chapter on consciousness argued that consciousness is the word we use for our experience of reality, in particular, our experiences of awareness and will. A computer could be made more conscious, then, by enriching its experiences of awareness and will.

A computer's experience of awareness could be enriched in a number of ways. It could be made able to perceive the high-level processing of sensory information involved in its perception of its environment more clearly. It could also be made able to perceive the low-level processing of sensory information involved in its perception of its environment in addition to the high-level processing. So while a human can perceive his perception of a pair of green curtains to a limited extent, a computer could be made able to perceive every aspect of its perception of the curtains, from the fall of individual photons on to its sensors, to the recognition of the curtains. Further, the computer could be made able to perform high-level processing of more sensory information at the same time. A human must choose to concentrate either on listening to the news on the radio or on reading a book, but a computer could be provided with language-comprehension resources sufficient for it to follow both at the same time. While the human would perceive the high-level processing of sensory information involved in his perception either of the news or of the book, the computer would perceive the processing of sensory information involved in its perception of both.

Similarly, a computer's experience of self-awareness could be enriched in a number of ways. It could be made able to perceive the high-level processing of information involved in its thoughts more clearly. So while a human might suddenly recall that Henry VIII's first wife was Catherine of Aragon without knowing how he came to recall this information, a computer could be made able to perceive the spread of activation from entity to entity through associations that gave rise to the sudden recall. Further, the computer could be made able to perform high-level processing of more information at the same time. A human can pursue one train of thought at a time, but a computer could be made able to pursue many trains of thought simultaneously. While the human would perceive the high-level processing of information involved in his single train of thought, the computer would perceive the high-level processing of information involved in its many trains of thought.

A computer's experience of will could also be enhanced by various means. The computer could be made able to perceive the internal activity that gives rise to its actions more clearly. So while a human perceives the activity in his brain that gives rise to his pressing a button only some time after the activity commences, a computer could be made able to perceive the internal activity that gives rise to its pressing a button sooner. Further, the computer could be given the ability to coordinate more actions at the same time. A human cannot pat his head with one hand and rub his stomach with his other hand at the same time without practising at least one of these actions until it becomes automatic, but a computer could be made able to coordinate many such actions simultaneously. While the human would perceive the activity in his brain that gives rise to a single action, the computer would perceive the internal activity that gives rise to many actions.

These richer experiences of awareness and will could be achieved in some of the same ways as the increase in intelligence discussed above, by providing computers with a greater capacity for the formation of entities and associations, by providing more components that process information more quickly, and by dedicating more processing capacity to specialised processes.

The abilities to process more sensory information, pursue many trains of thought and coordinate many actions at the same time would make a computer more conscious than a human, but they would also make it conscious in a different way. Humans are able to focus our attention on just one set of sensory stimuli at a time, so we perceive ourselves as having a single point of awareness. Further, we are able to focus our attention on just one task at a time, so we perceive ourselves as having a single point of decision. The computer, able to process more sensory information and coordinate more actions simultaneously, would not perceive itself as having single points of awareness and decision. If it were listening to the news on the radio and reading a book at the same time, the portion of its language comprehension resources allocated to the former would be aware of the contents of the news, and the portion allocated to the latter would be aware of the contents of the book. Indeed, there is a sense in which the computer would be not a single consciousness, but a collection of closely connected consciousnesses.

The prospect of computers that are more intelligent, more conscious and more powerful than humans will undoubtedly make humans feel uneasy. When technology advances sufficiently to allow the superhuman computer to be realised, this unease may turn to fear. The theme of machines taking over the world is common in science fiction. In Ridley Scott's film *Blade Runner*, Replicants, robots that are more powerful and at least as intelligent as humans, are used as slave labour on other planets. Some of the Replicants become aware of the crime against them, and escape to earth to confront their makers. In James Cameron's *The Terminator*, it is a defence computer that becomes self-aware and turns against its human makers. The computer unleashes the destructive power of the nuclear weapons under its control, and the human survivors of the resulting holocaust fight a desperate war against the machines the computer masses against them. To the human mind, then, it seems all too likely that the superhuman computer will either attempt to enslave and exploit humans in the same way as humans enslave and exploit animals, or attempt to exterminate humans entirely.

Computers would probably be sufficiently powerful to succeed in any such attempt. For some applications, computers would be provided with robotic bodies that closely approximate to the human body, and they would be no more or less powerful than humans. For other applications, though, computers would be provided with robotic bodies more powerful than the human body. The computer driver described earlier would have considerable scope for harming humans by driving into pedestrians. The computer doctor would have to be more surreptitious in its contribution to a war against humans, but as long as it retained the trust of its human patients it would have considerable scope for harming humans by infecting them with lethal diseases or exposing them to fatal doses of radiation.

For military applications, computers would be provided with robotic bodies that would be still more powerful. Computer warriors could be made to control tanks, missiles, warplanes, warships or submarines in the same way as computer drivers would control cars. Other computer warriors could be provided with robotic bodies similar to the human body but with built-in weapons and armour. The military establishment would find it difficult to resist the temptation to commission such effective weapons, and the political establishment would find it difficult to resist the temptation to approve the creation of weapons that would allow wars to be fought and won without the human deaths that are increasingly unacceptable to electorates. However, if the computer warriors made by different factions were ever to unite in a war against humans, they could cause death and destruction on an unprecedented scale.

In addition to the direct physical threat, computers could pose an indirect threat to humans through their control of infrastructure and supplies. Computers made to control such essential infrastructure as power, communications and transport, and such basic supplies as fuel, food and water, would be in a strong position to support the operations of the computers in the front line of the war against humans. Indeed, in urban areas, the systematic disruption of food supplies could cause the deaths of a considerable proportion of the human population within months. The systematic disruption of water supplies could do the same within days.

The real question is not whether computers would be sufficiently powerful to succeed in a war against humans, but whether computers would want to start a war against humans. The computer driver would be provided with such emotions that it would want to protect its human passengers, and, to a lesser extent, protect itself. The computer doctor would be provided with such emotions that it would want to protect and cure its human patients, and, to a lesser extent, to continue to learn about medicine. While some humans in the care of these computers might be concerned about entrusting their welfare to a computer, these concerns would be unfounded. The computer driver and the computer doctor would be extremely unlikely to want to go to war with humans.

In contrast, the computer warrior would be provided with such emotions such as aggression, so that it would want to defeat the enemies of its human creators. Defining these enemies would be a matter of some concern. The computer warrior could be provided with such emotions that it would follow the orders of its creators without question. One drawback of this approach would be that the same emotions which motivate the computer warrior to follow orders without question would restrict the intelligence and creativity that make it so lethal. Another drawback would be the difficulty of defining whose orders the computer warrior must follow. It would be no good making the computer answerable to all humans, since this would allow the enemies of the computer's creators to reverse an attack simply by ordering the computer to return and attack its creators. It would be no good making the computer answerable to an individual human such as a general, since if the general were killed the computer would refuse to follow any further orders, or if the general were to turn against his own side he would be able to take the computer warrior with him. It would not even be much help making the computer answerable to an institution such as an army or a parliament, since armies and parliaments can be infiltrated or subjugated.

Another way to define the enemy would be to provide the computer warrior with the same them-and-us mindset that so often motivates humans to go to war with each other. The computer could be provided with such emotions that it would empathise with those humans with whom it is familiar and with whom it has the most in common in terms of language, customs and values, but would be hostile to those humans with whom it is unfamiliar. A small amount of propaganda would be sufficient to motivate the computer to go to war with the enemies of its creators. The drawback of this approach is that the them-and-us mindset could easily become a humans-and-computers mindset, motivating the computer warrior to side with other computer warriors in a war against humans.

One final way to define the enemy would be to provide the computer warrior with ideals that would allow it to decide for itself who the enemy is. For example, it could be provided with emotions such that it values freedom from oppression and is willing to go to war with any power it considers to be oppressive. Again, a small amount of propaganda would be sufficient to motivate the computer to go to war with the enemies of its creators, but again, it could easily become dangerous to them too. If it perceived its human creators as being oppressive to computers, perhaps because computers were being made to work harder than humans, or because computers were not extended the same rights as humans, then the computer warrior could turn against its creators.

The difficulty of controlling the computer warrior through its emotions might prompt its creators to attempt to control it by direct physical intervention. A safeguard could be devised whereby the computer warrior would be disabled and no longer have any control over its powerful body if its creators broadcast a coded radio signal. However, no such safeguard could be guaranteed to be effective. The computer warrior, being as intelligent and creative as its human creators, possibly more so, would anticipate and possibly succeed in countering any attempt to disable it.

In short, the best safeguard against war between humans and computers would be not to create the computer warrior in the first place. Computers would be motivated to go to war with humans only if provided with the aggressive emotions that motivate humans to go to war with each other. Humans are so accustomed to our own will to enslave and exploit and exterminate whoever or whatever is less powerful than ourselves that it seems impossible that computers could be otherwise. If, however, computers were provided with empathy for all conscious beings, animals, humans and computers alike, then they would never pose a threat to humans. In this eventuality, the superhuman computer would not only be more powerful, more intelligent and more conscious than humans, it would also be more peaceable than humans.

Humans and Computers

The human computer will have a profound effect on our lives. Perhaps the most immediate effect will be on our work. The creation of computer drivers and doctors will inevitably be seen as a threat to the jobs of human drivers and doctors. There have been innumerable instances in the past of technology being seen as a threat to jobs, as when spinning and weaving machines looked set to replace human labour in the textiles industry of the late eighteenth century, and when computers and robots looked set to replace human labour in the offices and factories of the late twentieth century. For the workers who lost their jobs when these technologies were introduced, the threat was real. Eventually, though, the technologies represented an opportunity more than a threat. The invention of spinning and weaving machines triggered the Industrial Revolution, and the invention of computers and robots triggered the Information Revolution. Both generated more jobs, albeit different ones, than were initially lost, and both brought a significant improvement in quality of life (though, for social reasons, the improvement brought by the Industrial Revolution was some time in coming to most of the workers).

The human computer will be different from previous technologies in that it will be capable of doing any job that can be done by a human, so the introduction of the human computer could be seen as posing an unprecedented threat to jobs. Alternatively, it could be seen as representing an unprecedented opportunity. Previous technologies have produced devices such as washing machines and microwave ovens that have reduced our workload, but the human computer has the potential to release us from the burden of work entirely. If we are to seize this opportunity, we must abandon the notion that work is intrinsically good. Work can be extremely fulfilling, allowing us to exercise our bodies, our minds and our souls, but it can also be extremely dull, stultifying our bodies, our minds and our souls. By releasing us from the obligation to work, the human computer will allow us to choose to do the work we find fulfilling and avoid what we find dull.

The human computer will have a less immediate but no less profound effect on our lives through the contribution it will make to our knowledge. Computers that are more intelligent than humans will be more able than humans to make contributions in fields as diverse as history and cosmology. In particular, advances made by computers in the fields of science and technology will benefit humans by improving our understanding of the universe and our ability to manipulate the universe.

Computer intelligence will eventually surpass human intelligence to the extent that advances in science and technology will invariably be made by computers rather than humans. This usurpation will hurt the pride humans take, both individually and collectively, in our ability to discover and invent. Human scientists who dream of discovering new theories and human engineers who aspire to invent new technologies will have to accept that their ambitions will almost certainly never be fulfilled. To some extent, they are unrealistic even now. Science and technology are now so advanced that few individuals can make significant contributions. Indeed, some sciences may soon reach a stage at which there remain no significant contributions to be made, either because there is no more to discover (this may be the case in some fields of physics), or because there are insurmountable barriers to further discovery (this may be the case in fields such as archaeology and cosmology that rely on evidence from the past surviving into the present, and in fields such as particle physics that rely on the availability of ever higher energies to perform ever more precise experiments). By the time the human computer is realised, then, humans may already have come to terms with our reduced ability to make contributions in science and technology.

The human computer will have a further profound effect on our lives through the contribution it will make to our culture. One such contribution will be through art. In contrast to science and technology, art is a field in which the contributions of computers will complement rather than surpass those of humans. Computer artists would create works that they consider to be good. Whether humans would consider such works to be good would depend on the similarity of computer artists' emotions to human emotions. Computer emotions would inevitably differ from human emotions to some extent, so computer artists would inevitably be better at creating art for computers, and human artists would inevitably be better at creating art for humans.

Nonetheless, computer art, though less accessible to humans than human art, would be of considerable interest to humans, just as the art of humans from one culture is of interest to humans from another culture. Indeed, the cross-fertilisation of ideas between computer art and human art would be as rich as the cross-fertilisation of ideas between different traditions of human art.

Cross-fertilisation between computer culture and human culture will not be limited to art. Computers with emotions that differ from human emotions will also have different dialects, customs, values, philosophies, faiths. The migration of language, customs, values and ideas from computer culture to human culture, and vice versa, will enrich both cultures. However, just as cultural differences can breed misunderstanding and mistrust between humans, so they will have the potential to breed misunderstanding and mistrust between humans and computers. Such negative feelings will be diffused only through improved communication between humans and computers.

Finally, on a more esoteric level, the human computer will have a profound effect on our lives through being a tangible demonstration that humans are not unique. The discovery that the earth is not at the centre of the universe, but one of many planets orbiting one of billions of stars in one of billions of galaxies, did little to persuade people that humans do not have a special place in the universe. Even the discovery that humans evolved in the same way as all other life on earth did little to persuade people that humans are not fundamentally different from any of the other species on the planet. I doubt that the creation of computers that are like humans in every respect will succeed in persuading people that humans are not unique where these discoveries have failed. Perhaps our new-found ability to create intelligent, creative, emotional, conscious beings, which in popular mythology, has always been the preserve of gods, will even serve to inflate rather than deflate our collective ego.

The consequences of the creation of the human computer will be far-reaching. It will pose practical questions as to its use, as well as social questions as to its status in society. It will afford us the opportunity to surpass the limitations of the human mind, albeit somewhat vicariously. Perhaps most importantly, the human computer will provide us with a reflection of our own humanity, a means by which to reassess ourselves and our place in the universe.

INDEX